Nadia Nedjah, Leandro dos Santos Coelho and Luiza de Macedo Mourelle (Eds.)

# Quantum Inspired Intelligent Systems

# Studies in Computational Intelligence, Volume 121

Editor-in-chief
Prof. Janusz Kacprzyk
Systems Research Institute
Polish Academy of Sciences
ul. Newelska 6
01-447 Warsaw
Poland
*E-mail:* kacprzyk@ibspan.waw.pl

---

Further volumes of this series can be found on our homepage: springer.com

Vol. 100. Anthony Brabazon and Michael O'Neill (Eds.)
*Natural Computing in Computational Finance,* 2008
ISBN 978-3-540-77476-1

Vol. 101. Michael Granitzer, Mathias Lux and Marc Spaniol (Eds.)
*Multimedia Semantics - The Role of Metadata,* 2008
ISBN 978-3-540-77472-3

Vol. 102. Carlos Cotta, Simeon Reich, Robert Schaefer and Antoni Ligeza (Eds.)
*Knowledge-Driven Computing,* 2008
ISBN 978-3-540-77474-7

Vol. 103. Devendra K. Chaturvedi
*Soft Computing Techniques and its Applications in Electrical Engineering,* 2008
ISBN 978-3-540-77480-8

Vol. 104. Maria Virvou and Lakhmi C. Jain (Eds.)
*Intelligent Interactive Systems in Knowledge-Based Environment,* 2008
ISBN 978-3-540-77470-9

Vol. 105. Wolfgang Guenthner
*Enhancing Cognitive Assistance Systems with Inertial Measurement Units,* 2008
ISBN 978-3-540-76996-5

Vol. 106. Jacqueline Jarvis, Dennis Jarvis, Ralph Rönnquist and Lakhmi C. Jain (Eds.)
*Holonic Execution: A BDI Approach,* 2008
ISBN 978-3-540-77478-5

Vol. 107. Margarita Sordo, Sachin Vaidya and Lakhmi C. Jain (Eds.)
*Advanced Computational Intelligence Paradigms in Healthcare - 3,* 2008
ISBN 978-3-540-77661-1

Vol. 108. Vito Trianni
*Evolutionary Swarm Robotics,* 2008
ISBN 978-3-540-77611-6

Vol. 109. Panagiotis Chountas, Ilias Petrounias and Janusz Kacprzyk (Eds.)
*Intelligent Techniques and Tools for Novel System Architectures,* 2008
ISBN 978-3-540-77621-5

Vol. 110. Makoto Yokoo, Takayuki Ito, Minjie Zhang, Juhnyoung Lee and Tokuro Matsuo (Eds.)
*Electronic Commerce,* 2008
ISBN 978-3-540-77808-0

Vol. 111. David Elmakias (Ed.)
*New Computational Methods in Power System Reliability,* 2008
ISBN 978-3-540-77810-3

Vol. 112. Edgar N. Sanchez, Alma Y. Alanís and Alexander G. Loukianov
*Discrete-Time High Order Neural Control: Trained with Kalman Filtering,* 2008
ISBN 978-3-540-78288-9

Vol. 113. Gemma Bel-Enguix, M. Dolores Jiménez-López and Carlos Martín-Vide (Eds.)
*New Developments in Formal Languages and Applications,* 2008
ISBN 978-3-540-78290-2

Vol. 114. Christian Blum, Maria José Blesa Aguilera, Andrea Roli and Michael Sampels (Eds.)
*Hybrid Metaheuristics,* 2008
ISBN 978-3-540-78294-0

Vol. 115. John Fulcher and Lakhmi C. Jain (Eds.)
*Computational Intelligence: A Compendium,* 2008
ISBN 978-3-540-78292-6

Vol. 116. Ying Liu, Aixin Sun, Han Tong Loh, Wen Feng Lu and Ee-Peng Lim (Eds.)
*Advances of Computational Intelligence in Industrial Systems,* 2008
ISBN 978-3-540-78296-4

Vol. 117. Da Ruan, Frank Hardeman and Klaas van der Meer (Eds.)
*Intelligent Decision and Policy Making Support Systems,* 2008
ISBN 978-3-540-78306-0

Vol. 118. Tsau Young Lin, Ying Xie, Anita Wasilewska and Churn-Jung Liau (Eds.)
*Data Mining: Foundations and Practice,* 2008
ISBN 978-3-540-78487-6

Vol. 119. Slawomir Wiak, Andrzej Krawczyk and Ivo Dolezel (Eds.)
*Intelligent Computer Techniques in Applied Electromagnetics,* 2008
ISBN 978-3-540-78489-0

Vol. 120. George A. Tsihrintzis and Lakhmi C. Jain (Eds.)
*Multimedia Interactive Services in Intelligent Environments,* 2008
ISBN 978-3-540-78491-3

Vol. 121. Nadia Nedjah, Leandro dos Santos Coelho and Luiza de Macedo Mourelle (Eds.)
*Quantum Inspired Intelligent Systems,* 2008
ISBN 978-3-540-78531-6

Nadia Nedjah
Leandro dos Santos Coelho
Luiza de Macedo Mourelle
(Eds.)

# Quantum Inspired Intelligent Systems

With 60 Figures and 22 Tables

Springer

Nadia Nedjah
Universidade do Estado do Rio de Janeiro
Faculdade de Engenharia
sala 5022-D
Rua São Francisco Xavier 524
20550-900, MARACANÃ-RJ
Brazil
nadia@eng.uerj.br

Luiza de Macedo Mourelle
Universidade do Estado do Rio de Janeiro
Faculdade de Engenharia
sala 5022-D
Rua São Francisco Xavier 524
20550-900, MARACANÃ-RJ
Brazil
ldmm@eng.uerj.br

Leandro dos Santos Coelho
Pontifícia Universidade Católica do Paraná
Centro de Ciências Exatas e de Tecnologia
Rua Imaculada Conceição 1155
Curitiba-PR 80215-901
Brazil
leandro.coelho@pucpr.br

ISBN 978-3-642-09729-4 e-ISBN 978-3-540-78532-3

Studies in Computational Intelligence ISSN 1860-949X



Cover design: Deblik, Berlin, Germany

Printed on acid-free paper

9 8 7 6 5 4 3 2 1

springer.com

# Preface

Research on applying principles of quantum computing to improve the engineering of intelligent systems has been launched since late 1990s. This emergent research field concentrates on studying on quantum computing that is characterized by certain principles of quantum mechanics such as standing waves, interference, quantum bits, coherence, superposition of states, and concept of interference, combined with computational intelligence or soft computing approaches, such as artificial neural networks, fuzzy systems, evolutionary computing, swarm intelligence and hybrid soft computing methods. This volume offers a wide spectrum of research work developed using soft computing combined with quantum computing systems.

In Chapter 1, the authors present a novel Quantum-behaved PSO approach using Gaussian distribution (G-QPSO) to tune the design parameters of a Fuzzy Logic Control (FLC) system with PID (Proportional Integral Derivative) conception. The FLC-PID design is applied to a control valve with nonlinear dynamic behavior and numerical results indicate that proposed FLC-PID design with G-QPSO is effective for the control of reactor.

In Chapter 2, the authors propose to use quantum-inspired genetic algorithms for permutation flow shop scheduling, which is a typical NP-hard combinatorial optimization problem with strong engineering background. Simulations are carried out based on several single-objective and multi-objective benchmarks with some performance metrics and the results demonstrate the effectiveness of the proposed hybrid quantum-inspired genetic algorithms.

In Chapter 3, the authors explore a model for a quantum Hopfield artificial neural network, in which qubits are prepared in an initial state and allowed to evolve to steady state. Then, they derive a method for training the quantum network to converge to a stable target pattern, given an input pattern, and show that a spatial array of qubits can be trained to perform the CNOT, which with a rotation is a universal quantum gate.

In Chapter 4, the authors propose a new paradigm of intelligent systems and they apply it to engineering quantum intelligent mobile system through the fusion of quantum technology with mobile system. The authors show that this quantum intelligent mobile system has many potential applications in various areas and also offers a platform for the research on quantum or quantum-inspired technologies.

In Chapter 5, the authors show that qubit networks with long-range interactions governed by the Hebb rule can be used as quantum associative memories. They claim that the quantum associative memory model presented in this chapter is a first concrete example of the extension potential of the quantum information processing paradigm to "intelligent" tasks.

In Chapter 6, the authors propose a novel evolutionary algorithm for numerical optimization inspired by the multiple universes principle of quantum computing that presents faster convergence time for the benchmark problems. Results show that this algorithm can find better solutions, with less evaluations, when compared with similar algorithms, which greatly reduces the convergence time.

In Chapter 7, the authors illustrate how a quantum-inspired evolutionary algorithm (QIEA) using real number encodings can be constructed and examines the utility of the resulting algorithm on an important real-world problem. The author apply the proposed model to the calibration of an Option Pricing model. The results are robust but comparable to those of other algorithms.

We are very much grateful to the authors of this volume and to the reviewers for their tremendous service by critically reviewing the chapters. The editors would also like to thank Prof. Janusz Kacprzyk, the editor-in-chief of the Studies in Computational Intelligence Book Series and Dr. Thomas Ditzinger from Springer-Verlag, Germany for their editorial assistance and excellent collaboration to produce this scientific work. We hope that the reader will share our excitement on this volume and will find it useful.

*January 2008*

**Nadia Nedjah**, State University of Rio de Janeiro, Brazil
**Leandro S. Coelho**, Pontifical Catholic University of Parana, Brazil
**Luiza M. Mourelle**, State University of Rio de Janeiro, Brazil

# Contents

# List of Figures

# List of Tables

# 1

# Gaussian Quantum-Behaved Particle Swarm Optimization Applied to Fuzzy PID Controller Design

Leandro dos Santos Coelho[1], Nadia Nedjah[2], and Luiza de Macedo Mourelle[3]

[1] Production and Systems Engineering Graduate Program
Pontifical Catholic University of Parana, LAS/PPGEPS/PUCPR
Imaculada Conceicao, 1155, Zip code 80215-901, Curitiba, Parana, Brazil
leandro.coelho@pucpr.br

[2] Department of Electronics Engineering and Telecommunications,
Engineering Faculty,
State University of Rio de Janeiro,
Rua São Francisco Xavier, 524, Sala 5022-D,
Maracanã, Rio de Janeiro, Brazil
nadia@eng.uerj.br, http://www.eng.uerj.br/~nadia

[3] Department of System Engineering and Computation,
Engineering Faculty,
State University of Rio de Janeiro,
Rua São Francisco Xavier, 524, Sala 5022-D,
Maracanã, Rio de Janeiro, Brazil
ldmm@eng.uerj.br, http://www.eng.uerj.br/~ldmm

After having enjoyed an increasingly great popularity in Japan during the last two decades, Fuzzy Logic Control (FLC) systems have been investigated in many technical and industrial applications as a powerful modeling tool that can cope with the uncertainties and nonlinearities of modern control systems. Conventional control depends on the mathematical model of the plant being controlled. FLCs have become popular because they do not necessarily require a theoretical model of the plant which is to be controlled. The main drawback of these FLC methodologies in the industrial environment is the number of tuning parameters to be selected. Inspired by the classical particle swarm optimization (PSO) method and quantum mechanics theories, this work presents a novel Quantum-behaved PSO approach using Gaussian distribution (G-QPSO) to tune the design parameters of a FLC with PID (Proportional-Integral-Derivative) conception. The FLC-PID design has been applied to a control valve with nonlinear dynamic behavior. Numerical results

L. dos S. Coelho et al.: *Gaussian Quantum-Behaved Particle Swarm Optimization Applied to Fuzzy PID Controller Design*, Studies in Computational Intelligence (SCI) **121**, 1–15 (2008)
www.springerlink.com 

presented here indicate that proposed FLC-PID design with G-QPSO is effective for the control of reactor.

## 1.1 Introduction

Conventional PID (Proportional-Integral-Derivative) controllers are characterized with simple structure and simple design procedures. They enable good control performance and are therefore widely applied in industry [26]. More than 90% of all control loops are PID [1]. However, in a number of cases of nonlinear processes, such as those when parameter variations take place and/or when disturbances are present, control system based on a Fuzzy Logic Controller (FLC) may be a potential choice.

Fuzzy logic has the capability to handle imprecise information through linguistic expressions. Fuzzy logic techniques have been successfully applied on the control system without exact mathematical plant model in the past several years [17, 20, 39]. The fuzzy logic controllers (FLCs) are also shown to have satisfactory system performance for the complex systems [14, 43, 46].

FLCs implementation has been widely studied in the process control community. In the literature, many kinds of advanced FLCs are successfully designed and the efficiency of the fuzzy controller is tested, for example in issues of predictive control [2,52], adaptive control [7,23], sliding-mode control [4,32], and robust control [55]. Recently, a wide variety of fuzzy logic PID-like controllers (FLCs-PID) have been developed [3, 5, 14, 15, 19, 27, 30, 31, 35, 37, 41, 47, 48, 57].

Adding the difficulty of having to tune the membership functions, fuzzy control rules, and scaling factors of FLC-PID, to an unsystematic design procedure usually makes it impossible to achieve adequate performance because the controlled process is so complex. To overcome this flaw, a lot of research uses genetic algorithms [6, 8, 12, 36, 54] to be able to optimally set the parameters of the fuzzy controller.

In the last years, the fundamentals of quantum mechanics and physics have motivated the generation of optimization methods [18, 40]. Inspired by the particle swarm optimization (PSO) [13, 21] and quantum mechanics theories [38, 44], this work presents a novel Quantum-behaved PSO (QPSO) approach using Gaussian distribution (G-QPSO). A contribution of this chapter is the analysis and discussion of a FLC-PID issue using fuzzy PI and conventional D based on G-QPSO tuning of scaling factors. The FLC-PID designs using classical PSO, QPSO and G-QPSO methods are compared and have been applied to a nonlinear control valve.

The remaining sections of this chapter are organized as follows: in Section 1.2, fundamentals of FLCs and FLC-PID are shown. Section 1.3 describes the QPSO method using Gaussian distribution. In Section 1.4, the description of control valve and the simulation results of a strcture of FLC-PID are evaluated. Lastly, Section 1.5 presents our conclusions and future research work.

## 1.2 Fuzzy Control Algorithm

Fuzzy set theory was introduced by Zadeh in 1965 and can be utilized to transform an inexact knowledge into the form of a computer algorithm. In applications it is used by defining a fuzzy system with linguistic variables and with a set of if-then rules.

Being composed of control rules of conditional linguistic statements that state the relationship between input and output variables, the FLCs have the enticing abilities of emulating human knowledge and experiences and dealing with model uncertainty. FLCs are typically defined by the nonlinear mapping of system state space for the control space. However, it is possible to identify the result of a FLC as a control surface reflecting the operator's (or control engineer's) prior knowledge of the process. The control surface is coded into a knowledge base using a compiler to execute rule-base, termsets, and scale factors.

FLCs design presents a data flow that is dealt with a fuzzification phase, analysis and execution of rules and, a defuzzification phase. The fuzzification phase converts numeric variables (real numbers) into linguistic variables (fuzzy numbers), in other words, transforms the crisp data in the corresponding fuzzy set. The analysis and execution of rules (expressed by a set of production rules of the type IF <condition> THEN <conclusion>) are responsible for the evaluation of the rule-base. When a rule is activated, a calculation procedure is realized based on the previous values, before the output rule is attributed. The defuzzification phase is the reverse of the fuzzification phase, and converts linguistic variable values of the FLC output into crisp outputs [10]. The defuzzification phase describes the mapping of a fuzzy control action space in non-fuzzy control actions (see Fig. 1.1.

There are two major types of fuzzy controllers: FLCs based on linguistic model (Mamdani type) [34] and FLCs based on interpolative model (Takagi-Sugeno type) [51]. The major difference is that Mamdani fuzzy controllers use fuzzy sets whereas Takagi-Sugeno fuzzy controllers employ functions of input variables in consequent of fuzzy control rules. In this work, the Mamdani fuzzy controllers are adopted.

### 1.2.1 Fuzzy PID controller

Fuzzy logic approaches have been shown in numerous studies to be a simpler alternative to improve conventional PID control performance. Several FLC configurations have been evaluated and implemented in the processes control literature [14, 15, 19, 27, 35], where the usual FLC structure is the FLC-PI or the FLC-PD. This section presents the structures and design principles of some FLCs-PID, as alternative to the FLCs-PID complete design, which require three inputs, that will substantially expand the rule-base and make the design more difficulty. Then, the features of a FLC design based on fuzzy PI (Proportional-Integral) plus conventional D (FPI+D) are presented. The

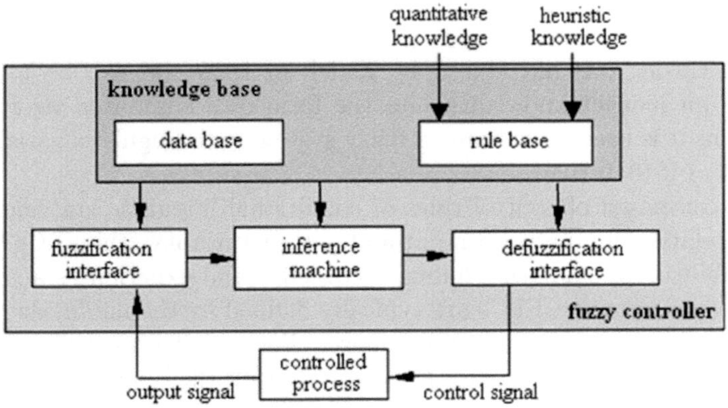


**Fig. 1.1.** Diagram of a linguistic FLC

FPI+D design, as proposed in [29,42], consists of a fuzzy PI plus the derivative control action of the process output, $U_D(k)$, defined as in (1.1)–(1.4).

$$u_{PID}(k) = K_c \cdot [u_{PI}(k) + u_D(k)] \tag{1.1}$$

$$u_{PI}(k) = K_e \cdot e(k) + K_{\Delta k} \cdot \Delta e(k) \tag{1.2}$$

$$e(k) = y(k) - k_r(k) \tag{1.3}$$

$$\Delta e(k) = e(k) - e(k-1) \tag{1.4}$$

wherein $K_c$, $K_e$, and $K_{\Delta e}$ are scaling factors, $e(k)$ is the error signal, $\Delta e(k)$ is the change of error signal, $y(k)$ is the output signal, and $y_r(k)$ is the setpoint signal. The variables $u_{PI}(k)$ and $u_D(k)$ are the control signals of PI and D controller, respectively. In this context, the derivative control action is defined as in (1.5), where $K_D$ is the derivative gain constant.

$$u_D(k) = K_D \cdot y(k) - y_r(k) \tag{1.5}$$

The principle of this design is that the FPI+D control algorithm has the benefit of implementing the derivative control on the output, avoiding derivative kicks for step setpoint (reference) changes. The basic control diagram and membership functions of FPI+D are shown in Fig. 1.2 and Fig. 1.3, respectively. Throughout this chapter, the following linguistic terms are used: P (Positive), N (Negative), Z (Zero), NS (Negative Small), NB (Negative Big), PS (Positive Small) and PB (Positive Big), and the triangular shape and singleton forms are adopted how membership functions of the FPI+D design.

## 1.3 Quantum-behaved Particle Swarm Optimization

In terms of classical mechanics, a particle is depicted by its position vector $x$ and velocity vector $v$, which determine the trajectory of the particle. The

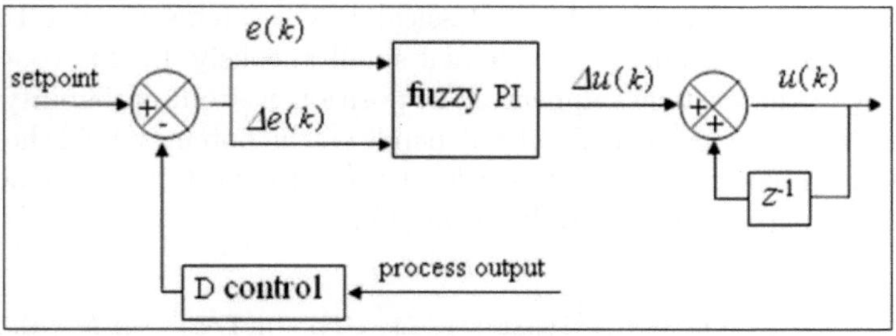


**Fig. 1.2.** Basic control diagram of the FPI+D

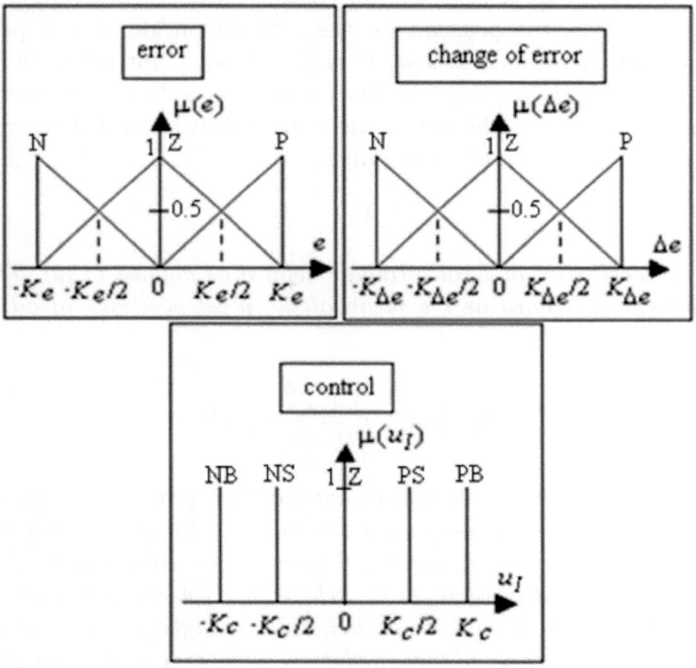


**Fig. 1.3.** Membership functions of the FPI+D

particle moves along a determined trajectory in Newtonian mechanics, but this is not the case in quantum mechanics. In quantum world, the term trajectory is meaningless because $x$ and $v$ of a particle can not be determined simultaneously according to uncertainty principle. Therefore, if individual particles in a PSO system have quantum behavior, the PSO algorithm is bound to work in a different fashion [49, 50].

In the quantum model of a PSO called here QPSO, the state of a particle is depicted by wavefunction $\psi(x,t)$ (Schrödinger equation) [28], instead of position and velocity. The dynamic behavior of the particle is widely divergent form that of that the particle in classical PSO systems in that the exact values of $x$ and $v$ cannot be determined simultaneously. In this context, the probability of the particle's appearing in position $x_i$ from probability density function $|\psi(x,t)|^2$, the form of which depends on the potential field the particle lies in [33]. Employing the Monte Carlo method, the particles move according to iterative equation (1.6) introduced in [49]:

$$x_{i,j}(t+1) = \begin{cases} p_i(t) + \beta \cdot |Mbest_j(t) - x_{i,j}(t)| \cdot \ln(1/u) & \text{if } k \geq 0.5 \\ p_i(t) - \beta \cdot |Mbest_j(t) - x_{i,j}(t)| \cdot \ln(1/u) & \text{if } k < 0.5 \end{cases} \quad (1.6)$$

where $x_{i,j}(t+1)$ is the position for the $j$-th dimension of $i$-th particle in $t$-th generation (iteration), $Mbest_j(t)$ is the global point called *mainstream thought* or *mean best) for the j-th dimension; $\beta$ is a design parameter called contraction-expansion coefficient, u and k are values generated according to a uniform probability distribution in range*

$$0, 1$$

*and $p_i(t)$ is local point or local attractor defined as defined in [9]. The mainstream thought is defined as the mean of the pbest positions of all particles and it given in (1.7).*

$$Mbest_j(t) = \frac{1}{N} \sum_{j=1}^{N} p_{g,j}(t) \quad (1.7)$$

*where N represents the dimension of optimization problem, g represents the index of the best particle among all the particles of the swarm in j-th dimension. In this case, the local attractor [9] to guarantee convergence of the PSO presents coordinates defined in (1.8), where $p_{k,i}$ (pbest) represents the best previous i-th position of the k-th particle and $p_{g,i}$ (gbest) represents the i-th position of the best particle of the population. Positive constants $c_1$ and $c_2$ are the cognitive and social components, respectively, which are the acceleration constants responsible for varying the particle velocity towards pbest and gbest, respectively.*

$$p_i(t) = \frac{c_1 \cdot p_{k,i} + c_2 \cdot p_{gi}}{C_1 + c_2} \quad (1.8)$$

*The procedure for implementing the QPSO is given by the following steps. It appeared first in [11]:*

1. *Initialization of swarm positions: Initialize a population (array) of particles with random positions in the n dimensional problem space using a uniform probability distribution function.*
2. *Evaluation of particle's fitness: Evaluate the fitness value of each particle.*

3. *Comparison of each particle's fitness with its pbest (personal best): Compare each particle's fitness with the particle's pbest. If the current value is better than pbest, then set a novel pbest value equals to the current value and the pbest location equals to the current location in $n$-dimensional space.*
4. *Comparison of each particle's fitness with its gbest (global best): Compare the fitness with the population's overall previous best. If the current value is better than gbest, then reset gbest to the current particle's array index and value.*
5. *Updating of global point: Calculate the Mbest using equation (7).*
6. *Updating of particle's position: Change the position of the particles where $c_1$ and $c_2$ are two random numbers generated using a uniform probability distribution in the range*

$$0, 1$$

.

7. *Repeating the evolutionary cycle: Loop to Step 2 until a stop criterion is met, usually a sufficiently good fitness or a maximum number of iterations (generations).*

### 1.3.1 QPSO using Gaussian distribution

*Various approaches using Gaussian, Cauchy and exponential probability distributions to generate random numbers to updating the velocity equation of classical PSO have been proposed [22, 24, 25, 45]. Following the same line of study, we present novel operator in QPSO using random numbers based on Gaussian probability distribution. The design of methods to improve the convergence of QPSO is a challenging issue in the continuous optimization applications. A novel QPSO approach based on Gaussian distribution is proposed here.*

*Generating random numbers using Gaussian distribution sequences with zero mean and unit variance for the stochastic coefficients of QPSO may provide a good compromise between the probability of having a large number of small amplitudes around the current points (fine tuning) and a small probability of having higher amplitudes, which may allow particles to move away from the current point and escape from local minima.*

*In this work, firstly, random numbers are generated using the absolute value* $|\cdot|$ *of the Gaussian probability distribution with zero mean and unit variance, i.e.,* $|N(0,1)|$*, and then mapped to a truncated signal given by* $G = 0.33 \cdot |N(0,1)|$*. The QPSO approach combined with Gaussian mutation operator (G-QPSO) is described as in (1.9). In this case, the parameter* $\beta$ *of equation (1.6) is modified by the equation (1.9).*

$$x_{i,j}(t+1) = \begin{cases} p_i(t) + G \cdot |Mbest_j(t) - x_{i,j}(t)| \cdot \ln(1/u) & if\, k \geq 0.5 \\ \\ p_i(t) - G \cdot |Mbest_j(t) - x_{i,j}(t)| \cdot \ln(1/u) & if\, k < 0.5 \end{cases} \tag{1.9}$$

## 1.4 Case Study and Simulation Results

### 1.4.1 Description of Control Valve

*The case study is a control valve. The control valve is an opening with adjustable area. Normally it consists of an actuator, a valve body and a valve plug. The actuator is a device that transforms the control signal to movement of the stem and valve plug. We picked the model from [53] in which the control valve is described by a Wiener model (the nonlinear element follows linear block). The Wiener model is given in (1.10) and (1.11:*

$$x(t) = 1.5714 \cdot x(t-1) + 0.6873 \cdot x(t-2) + 0.0616 \cdot u(t-1) + 0.0543 \cdot u(t-2) \quad (1.10)$$

$$y(t) = f_n(x(t)) = \frac{x(t)}{\sqrt{0.10 + 0.90 \cdot (x(t))^2}} \quad (1.11)$$

where $u(t)$ is the control pressure, $x(t)$ is the stem position, and $y(t)$ is the flow through the valve which is the controlled variable. The nonlinear behavior of the control valve described by equation (1.11) is shown in Fig. 1.4. The input to the process, $u(t)$, is constrained between

$$0, 1$$

.

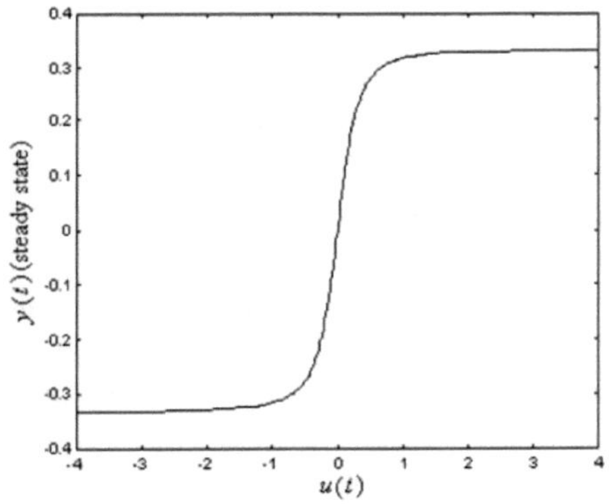


**Fig. 1.4.** Static curve of control valve

### 1.4.2 Simulation results

The effectiveness and advantages of the proposed FPI+D based on G-QPSO is demonstrated through controlling a valve control, where the following setup is adopted:

- Reference trajectory in optimization phase (servo behavior): The desired reference signal is given by $yr(t) = 0.7$ (from sample 1 to 100), $yr(t) = 0.3$ (from sample 101 to 200), $yr(t) = 0.8$ (from sample 201 to 300), $yr(t) = 0.4$ (from sample 301 to 400), $yr(t) = 0.6$ (from sample 401 to 500) and $yr(t) = 0.5$ (from sample 501 to 600).
- Optimization procedure: The PSO approaches is used in optimization procedure of scaling factor parameters of FPI+D. In the sequel it illustrates the main features of the PSO approaches employed:
  1. Objective function: In this work, the objective function to be minimized is given in (1.12):

$$y(t) = f_n(x(t)) = \sum_{k=1}^{N} |e(k)| \tag{1.12}$$

  2. Search space used in PSO approaches: $K_e$, $K_{\Delta e}$, $K_c \in [0, 2]$ and $K_D \in [-0.5, 0.5]$.
  3. Stopping criterion: The adopted termination criterion is the number of generations is equal to 20. The adopted population size is 20. In this case, the number of experiments (runs) for each design is set to 30.

PSO, QPSO and Q-QPSO methods were implemented using Matlab 6.5 to run on a PC compatible with Pentium IV, a 3.2 GHz processor and 2 GB of RAM. A total of 600 cost function evaluations were made with each optimization approach in each run. In this work, the optimization approaches use $c_1 = c_2 = 2.05$. The QPSO uses $\beta$ with a linear reduction equation with initial and final values of 1.0 and 0.4. In this case study, the adjustment of the rule base of FPI+D was accomplished by fine tuning and heuristic corrections linked to the knowledge of the process to be controlled, as shown in Fig. 1.5.

| $e$ \ $\Delta e$ | N | Z | P |
|---|---|---|---|
| N | NS | NS | NB |
| Z | Z | Z | Z |
| P | PB | PS | PS |

**Fig. 1.5.** Rule base adopted for the FPI+D design

Table 1.1 and Table 1.2 summarize the performance and design parameters of FPI+D optimized by classical PSO [13], QPSO [50], and G-QPSO methods for 30 runs. As can be seen, for the control valve, the best mean, minimum, and maximum from the 30 runs performed was using G-QPSO. In this context, the best solution obtained using G-QPSO with $J = 17.8572$. Servo simulation results of the FPI+D using G-QPSO for the control valve are shown in Fig. 1.6. However, G-QPSO reached solutions closer to the best solution of QPSO approach.

**Table 1.1.** Simulation results for FPI+D using data of best objective function in each run – Best Design

| Optimization method | Mean time (in sec) | Best design based on J | | | |
|---|---|---|---|---|---|
| | | $K_e$ | $K_{\Delta e}$ | $K_c$ | $K_D$ |
| PSO | 119.89 | 0.9912 | 1.9931 | 0.2005 | -0.4912 |
| QPSO | 120.15 | 0.9850 | 1.9920 | 0.3326 | -0.2801 |
| G-QPSO | 120.86 | 0.6773 | 1.9999 | 0.2386 | -0.4814 |

**Table 1.2.** Simulation results for FPI+D using data of best objective function in each run – Objective function

| Optimization method | Mean time (in sec) | objective function, J | | | |
|---|---|---|---|---|---|
| | | max | mean | min | std. deviation |
| PSO | 119.89 | 30.5689 | 22.7215 | 18.5067 | 3.7965 |
| QPSO | 120.15 | 20.0168 | 18.9971 | 18.5008 | 0.5018 |
| G-QPSO | 120.86 | 17.9393 | 17.8842 | 17.8572 | 0.0258 |

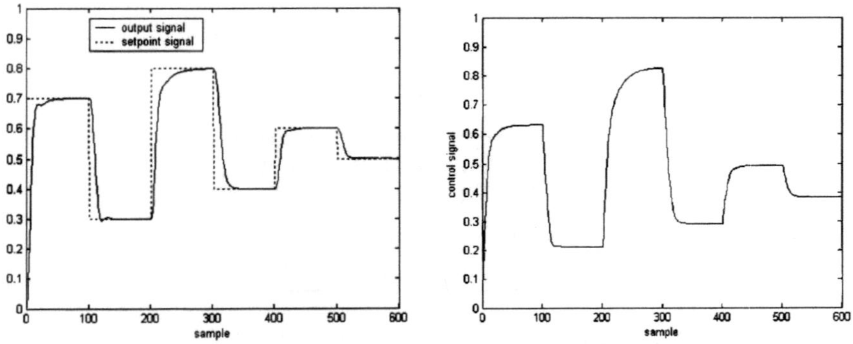


(a) (b)

**Fig. 1.6.** Best results of servo behavior using FPI+D with G-QPSO

In this study, the analysis of regulatory behavior of FPI+D is based on the rejection of additive disturbances and parametric changes in control valve. The test of regulatory behavior is based on the rejection of additive disturbances in the process output when: *(i)* sample 70 with $y(t) = y(t) + 0.2$; *(ii)* sample 150 with $y(t) = y(t) - 0.2$; *(iii)* sample 260 $y(t) = y(t) - 0.2$; *(iv)* sample 360 with $y(t) = y(t) - 0.2$; *(v)* sample 440 with $y(t) = y(t) + 0.3$ and *(vi)* sample 530 with $y(t) = y(t) - 0.4$. Simulation results of the FPI+D using QPSO based on the rejection of additive disturbances are shown in Fig. 1.7.

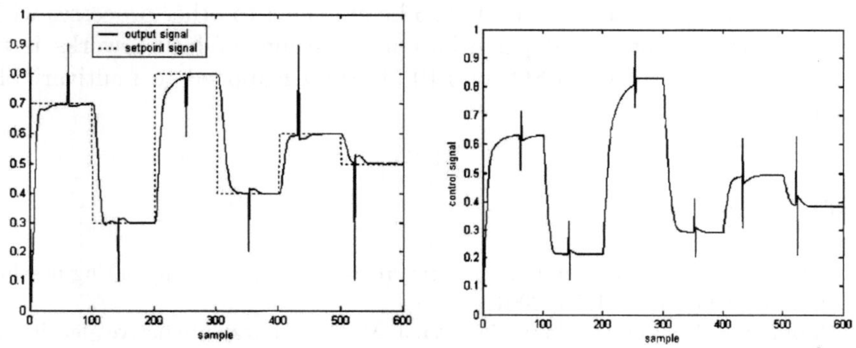


(a) (b)

**Fig. 1.7.** Best result of regulatory behavior using FPI+D with G-QPSO

Performance of FPI+D design was affected by nonlinearity of control valve. Furthermore, the FPI+D design based on G-QPSO obtained fast response, reasonable control activity, and good setpoint tracking and rejection of disturbances abilities. The good performance indicated by the FPI+D using G-QPSO approach confirms the usefulness and robustness of the proposed method for practical applications.

## 1.5 Conclusion and Future Work

FLCs have been developed successfully for the analysis and control of nonlinear systems. Although FLCs-PID have achieved practical success in controlling plants that are mathematically poorly modeled and where experienced operators are available, it is sometimes difficult to specify the rule base for some plants, or the need may arise to tune the parameters if the plant changes. With this motivation several mathematical methods to tuning and optimization of design parameters of FLCs-PID have been proposed in literature, as mentioned in Section 1.1.

This paper presented the development of a G-QPSO optimization method to the FPI+D design. In the paper, the effectiveness of the proposed control schemes is shown in simulations of a control valve. The utilization of PSO, QPSO, and G-QPSO approaches avoids the tedious manual trial-and-error procedure and it presents robustness in tuning of FPI+D design parameters.

The G-QPSO method proposed in this paper presents some promising features. Parametric uncertainties affect the closed loop system dynamics. However, the effectiveness of the proposed FPI+D strategy using G-QPSO is corroborated in Fig. 1.6 and Fig. 1.7. The simulations suggest that a typical nonlinear process such as the control valve can be effectively controlled by the proposed FPI+D with fast setpoint tracking and good disturbance rejection.

The proposed method is expected to be extended to other processes with parameter uncertainties and perturbations. The aim of future works is to investigate the use of G-QPSO for FPI+D tuning applied to multivariable chemical processes.

## References


1. Äström, K. J. and Hägglund, T., The future of PID control, Control Engineering Practice 9(11), 1163–1175, 2001.
2. Baptista, L. F., Sousa, J. M. and Costa, J. M. G. S., Fuzzy predictive algorithms applied to real-time force control, Control Engineering Practice 9(4), 411–423, 2001.
3. Carvajal, J., Chen, G. and Ogmen, H., Fuzzy PID controller: design, performance, evaluation and stability analysis, Information Sciences 123(3–4), 249–270, 2000.
4. Chang, Y. C., Adaptive fuzzy-based tracking control for nonlinear SISO systems via VSS H8 approaches, IEEE Transactions on Fuzzy Systems 9(2), 278–292, 2001.
5. Chen, G., Conventional and fuzzy PID controllers: an overview, International Journal of Intelligent Control and Systems 1(2), 235–246, 1996.
6. Cho, H. -J., Cho, K. -B. and Wang B. -H., Fuzzy-PID hybrid control: automatic rule generation using genetic algorithms, Fuzzy Sets and Systems 92(3), 305–316, 1997.
7. Cho, Y. -W., Park, C. -W. and Park, M., An indirect model reference adaptive fuzzy control for SISO Takagi-Sugeno model, Fuzzy Sets and Systems 131(2), 197–215, 2002.
8. Chou, C. -H., Genetic algorithm-based optimal fuzzy controller design in the linguistic space, IEEE Transactions on Fuzzy Systems 14(3), 372–395, 2006.
9. Clerc, M. and Kennedy, J. F., The particle swarm: explosion, stability and convergence in a multi-dimensional complex space, IEEE Transactions on Evolutionary Computation 6(1), 58–73, 2002.
10. Coelho L.S. and Coelho A. A. R., Fuzzy PID controllers: structures, design principles and application for nonlinear practical process, Advances in Soft Computing–Engineering Design and Manufacturing, Roy, R., Furushashi, T. and Chawdhry, K. (eds.), Springer, London, UK, 147–159, 1999.
11. Coelho, L. S., A quantum particle swarm optimizer with chaotic mutation operator, Chaos, Solitons and Fractals (in press), 2007.

12. Cordón, O., Gomide, F., Herrera, F., Hoffmann, F. and Magdalena, Ten years of genetic fuzzy systems: current framework and new trends, Fuzzy Sets and Systems 141(1), 5–31, 2004.
13. Eberhart R.C. and Kennedy J. F., A new optimizer using particle swarm theory, Proceedings of International Symposium on Micro Machine and Human Science, Japan, 39–43, 1995.
14. Feng, G., A survey on analysis and design of model-based fuzzy control systems, IEEE Transactions on Fuzz Systems 14(5), 676–697, 2006
15. Golob, M., Decomposed fuzzy proportional-integral-derivative controllers, Applied Soft Computing 1(3), 201–214, 2001
16. Higashi, N. and Iba, H., Particle swarm optimization with gaussian mutation. Proceedings of the IEEE Swarm Intelligence Symposium, Indianapolis, IN, USA, 72–79, 2003.
17. Hirota, K. and Sugeno, M., Industrial Applications of Fuzzy Technology in the World, Advances in Fuzzy Systems: Applications and Theory, vol. 2, Singapore, World Scientific, 1995.
18. Hogg, T. and Portnov, D. S., Quantum optimization, Information Sciences 128(3–4), 181–197, 2000.
19. Hu, B., Mann, G. K. I. and Gosine, R. G., New methodology for analytical and optimal design of fuzzy PID controllers, IEEE Transactions on Fuzzy Systems 7(5), 521–539, 1999.
20. Isermann, R. On fuzzy logic applications for automatic control, supervision and fault diagnosis, IEEE Transactions on Systems, Man and Cybernetics – Part B: Cybernetics 28(2), 221–235, 1998.
21. Kennedy, J. F. and Eberhart, R.C., Particle swarm optimization, Proceedings of IEEE International Conference on Neural Networks, Perth, Australia, 1942–1948, 1995.
22. Kennedy, J. F., Bare bones particle swarms, Proceedings of the IEEE Swarm Intelligence Symposium, Indianapolis, IN, USA, 80–87, 2003.
23. Koo, K. -M., Stable adaptive fuzzy controller with time-varying dead-zone, Fuzzy Sets and Systems 121(1), 161–168, 2001.
24. Krohling, R. and Coelho, L. S. Coevolutionary particle swarm optimization using Gaussian distribution for solving constrained optimization problems, IEEE Transactions on Systems, Man and Cybernetics, Part B: Cybernetics 36(6), 1407–1416, 2006
25. Krohling, R., Coelho, L. S. PSO-E: particle swarm with exponential distribution, IEEE World Congress on Computational Intelligence, Proceedings of IEEE Congress on Evolutionary Computation, Vancouver, Canada, 5577–5582, 2006.
26. Kukolj, D. D., Kuzmanovic, S. B. and Levi E., Design of a PID-like compound fuzzy logic controller, Engineering Applications of Artificial Intelligence 14(6), 785–803, 2001.
27. Kwok, D. P., Tam, P., Li, C. K. and Wang, P., Linguistic PID controllers, Proceedings of 11th World Congress of IFAC, Tallin, Estonia, USSR, vol. 7, 192–197, 1990.
28. Levin, F. S., An introduction to quantum theory, Cambridge University Press, 2002.
29. Li, H. X. and Gatland, H. B., Enhanced methods of fuzzy logic control. Proceedings of FUZZ-IEEE/IFES, vol. 1, Yokohama, Japan, 331–336, 1995.

30. Li, H. X., Zhang, L., Cai, K. Y. and Chen, G., An improved robust fuzzy-PID controller with optimal fuzzy reasoning, IEEE Transactions on Systems, Man and Cybernetics – Part B: Cybernetics 35(6), 1283–1294, 2005.
31. Li, Y. and Ng, K. C., Reduced rule-base and direct implementation of fuzzy logic control, Proceedings of 13th World Congress of IFAC, San Francisco, CA, USA, 85–90, 1997.
32. Lin, J. -M., Huang, S. -J. and Shin, K. -R., Application of sliding surface-enhanced fuzzy control for dynamic state estimation of a power system, IEEE Transactions on Power Systems 18(2), 570–577, 2003.
33. Liu, J., Xu, W. and Sun, J., Quantum-behaved particle swarm optimization with mutation operator, Proceedings of 17th International Conference on Tools with Artificial Intelligence, Hong Kong, China, 2005.
34. Mamdani, E. and Assilian, S., An experiment in linguistic synthesis with a fuzzy logic controller, International Journal on Man Machine Studies 7(1), 1–13, 1975.
35. Mann, G. K. I., Hu, B. -G. and Gosine, R. G., Analysis of direct action fuzzy PID controller structures, IEEE Transactions on Systems, Man and Cybernetics - Part B: Cybernetics 29(3), 371–388, 1999.
36. Marseguerra, M., Zio, E. and Cadini, F., Genetic algorithm optimization of a model-free fuzzy control system, Annals of Nuclear Energy 32(7), 712–728, 2005.
37. Misir, D., Malki, H. A. and Chen, G., Design and analysis of a fuzzy proportional-integral-derivative controller, Fuzzy Sets and Systems 79(3), 297–314, 1996.
38. Pang, X. F., Quantum mechanics in nonlinear systems. World Scientific Publishing Company, River Edge, NJ, USA, 2005.
39. Passino, K. M. and S. Yurkovich, Fuzzy Control, Addison Wesley, Menlo Park, CA, USA, 1998.
40. Protopescu, V. and Barhen, J., Solving a class of continuous global optimization problems using quantum algorithms, Physics Letters A 296, 9–14, 2002.
41. Qiao, W. Z. and Mizumoto, M., PID type fuzzy controller and parameters adaptive method, Fuzzy Sets and Systems 78(1), 23–35, 1996.
42. Qin, S. J., Auto-tuned fuzzy logic control, Proceedings of the American Control Conference, Baltimore, Maryland, USA, 2465–2469, 1994.
43. Sala, A., Guerra, T. M. and Babuska, R., Perspectives of fuzzy systems and control, Fuzzy Sets and Systems 156(3), 432–444, 2005.
44. Schweizer, W., Numerical quantum dynamics, Hingham, MA, USA, 2001.
45. Secrest, B. R. and Lamont, G. B., Visualizing particle swarm optimization – gaussian particle swarm optimization, Proceedings of the IEEE Swarm Intelligence Symposium, Indianapolis, IN, USA, 198–204, 2003.
46. Serra, G. L. O. and Bottura, C. P., Multiobjective evolution based fuzzy PI controller design for nonlinear systems, Engineering Applications of Artificial Intelligence 19(2), 157–167, 2006.
47. Shayeghi, H., Shayanfar, H. A. and Jalili, A., Multi-stage fuzzy PID power system automatic generation controller in deregulared environments, Energy Conversion and Management 47(18–19), 2829–2845, 2006.
48. Sio, K. C. and Lee, C. K., Stability of fuzzy PID controllers, IEEE Transactions on Systems, Man and Cybernetics – Part A: Systems and Humans 28(4), 490–495, 1998.
49. Sun J., Feng, B. and Xu, W., Particle swarm optimization with particles having quantum behavior, Proceedings of Congress on Evolutionary Computation, Portland, Oregon, USA, 325–331, 2004.

50. Sun, J., Xu, W. and Feng, B., Adaptive parameter control for quantum-behaved particle swarm optimization on individual level, Proceedings of IEEE International Conference on Systems, Man and Cybernetics, Big Island, HI, USA, 3049–3054, 2005.
51. Takagi, T. and Sugeno, M. (1985). Fuzzy identification of systems and its applications to modeling and control, IEEE Transactions on Systems, Man and Cybernetics 15, 116–132.
52. Vukovic, P. D., One-step ahead predictive fuzzy controller, Fuzzy Sets and Systems 122(1), 107–115, 2001.
53. Wigren, T., Recursive prediction error identification using the nonlinear Wiener model, Automatica 29(4), 1011–1025, 1993.
54. Wu, C. -J. and Liu, G. -Y., A genetic approach for simultaneous design of membership functions and fuzzy control rules, Journal of Intelligent and Robotic Systems 28(3), 195–211, 2000.
55. Yang, Y., Direct robust adaptive fuzzy control (DRAFC) for uncertain nonlinear systems using small gain theorem, Fuzzy Sets and Systems 151(1), 79–97, 2005.
56. Zadeh, L. A., Fuzzy sets, Information and Control 8, 338–353, 1996.
57. Zhao, Z. Y., Tomizuka, M. and Isaka, S., Fuzzy gain scheduling of PID controllers, IEEE Transactions on Systems, Man and Cybernetics 23(5), 1392–1398, 1993.

# 2

# Quantum-inspired genetic algorithms for flow shop scheduling

Ling Wang[1] and Bin-bin Li[2]

[1] Department of Automation, Tsinghua University, Beijing 100084, China
wangling@mail.tsinghua.edu.cn

[2] Center for Information and Systems Engineering (CISE) and Department of Manufacturing Engineering Boston University, 15 St. Mary's Street, Brookline, MA 02446
libinbin@bu.edu

In this chapter, quantum-inspired genetic algorithms are proposed for permutation flow shop scheduling, which is a typical NP-hard combinatorial optimization problem with strong engineering background. For the single-objective scheduling problems, a hybrid quantum-inspired genetic algorithm (HQGA) is developed to achieve more satisfactory performance. In the HQGA, a quantum-inspired GA (QGA) based on Q-bit representation is applied for exploration in discrete 0-1 hyperspace by using updating operator of quantum gate as well as genetic operators of Q-bit. And the random key representation is used to convert the Q-bit representation to job permutation for evaluating the objective value of the schedule solution. Moreover, a permutation-based GA (PGA) is applied for both performing exploration in permutation-based scheduling space and stressing exploitation for good schedule solutions. The HQGA can be viewed as a fusion of micro-space based search (Q-bit based search) and macro-space based search (permutation-based search). For the multi-objective scheduling problems, some special strategies are utilized on the basis of the HQGA for single-objective scheduling. In particular, the randomly weighted linear sum function is used in the QGA to evaluate solutions in multi-objective sense, and a non-dominated sorting technique including classification of Pareto front and fitness assignment is applied in the PGA regarding to both proximity and diversity. In addition, two trimming techniques for population are proposed and applied in the HQGA to maintain diversity of population. Simulations are carried out based on several single-objective and multi-objective benchmarks with some performance metrics. The results and comparisons demonstrate the effectiveness of the proposed hybrid quantum-inspired genetic algorithms.

L. Wang and B.-B. Li: *Quantum-inspired genetic algorithms for flow shop scheduling*, Studies in Computational Intelligence (SCI) **121**, 17–56 (2008)
www.springerlink.com 

## 2.1 Introduction

Quantum computing is a research area that includes concepts like quantum mechanical computers and quantum algorithms. So far, many efforts on quantum computer have progressed actively due to its superiority to classical computer on various specialized problems. There are well-known quantum algorithms such as Grover's database search algorithm [1] and Shor's quantum factoring algorithm [2]. During the past two decades, evolutionary algorithms have gained much attention and wide applications, which are essentially stochastic search methods based on the principles of natural biological evolution [3]. Since later 1990s, research on merging evolutionary computing and quantum computing has been started and gained attention both in physics, mathematics and computer science fields. One of the important topics concentrates on quantum-inspired evolutionary computing characterized by certain principles of quantum mechanisms for a classic computer [4–7].

Recently, some quantum-inspired genetic algorithms (QGAs) have been proposed for some combinatorial optimization problems, such as traveling salesman problem [4] and knapsack problem [5–7]. However, to the best of our knowledge, there is litter published research work on QGA for scheduling problems. As we know, solution of flow shop scheduling [8] is often not a string of 0-1 but a permutation of all jobs. Thus, the previous QGA cannot directly apply to scheduling problems. Obviously, it is a challenge to employ the algorithm for different problems other than those that inventors originally focused on. In this chapter, we propose some hybrid quantum-inspired genetic algorithms for single-objective and multi-objective permutation flow shop scheduling (PFSS). For the single-objective scheduling problems, a QGA based on Q-bit representation is applied for exploration in discrete 0-1 hyperspace by using updating operator of quantum gate as well as genetic operators of Q-bit, and the random key representation is used to convert the Q-bit representation to job permutation for evaluating the objective value of the schedule solution. Moreover, a permutation-based GA (PGA) is applied for both performing exploration in permutation-based scheduling space and stressing exploitation for good schedule solutions. For the multi-objective scheduling problems, the randomly weighted linear sum function is used in the QGA to evaluate solutions in multi-objective sense, and a non-dominated sorting technique including classification of Pareto front and fitness assignment is applied in the PGA regarding to both proximity and diversity. In addition, two trimming techniques for population are proposed to maintain diversity of population. Simulations are carried out based on several single-objective and multi-objective benchmarks, and comparisons are performed with several performance metrics for multi-objective scheduling, and the effectiveness of the proposed hybrid quantum-inspired genetic algorithms is demonstrated.

The organization of the remaining content is as follow. In Sect. 2.2 and Sect. 2.3, the PFSS and the QGA are reviewed, respectively. In Sect. 2.4, the hybrid QGA for single-objective FSSP is presented, and the hybrid QGA for multi-objective FSSP is presented in Sect. 2.5. In Sect. 2.6, some performance metrics about multi-objective optimization are introduced. Then, simulation results and comparisons for single-objective and multi-objective FSSP are provided in Sect. 2.7 and Sect. 2.8, respectively. Finally, we end the paper with some conclusions and future work in Sect. 2.9.

## 2.2 Permutation flow shop scheduling

### 2.2.1 Problem statement

Flow shop scheduling problem (FSSP) is a class of widely studied scheduling problem with strong engineering background, which represents nearly a quarter of manufacturing systems, assembly lines and information service facilities nowadays, and has earned a reputation for being a typical strongly NP-complete combinatorial optimization problem [9]. So far, FSSP has attracted much attention and wide research in both Computer Science and Operation Research fields.

As a simplification of FSSP, the permutation flow shop scheduling (PFSS) with $J$ jobs and $M$ machines is commonly defined as follows [8]. Each of $J$ jobs is to be sequentially processed on machine $1, \ldots, M$ . The processing time $p_{i,j}$ of job $i$ on machine $j$ is given. At any time, each machine can process at most one job and each job can be processed on at most one machine. The sequence in which the jobs are to be processed is the same for each machine. The objective is to find a permutation of jobs or a permutation set to minimize one or more criteria.

For the PFSS problems, we denote $S_i(\sigma)$ and $C_i(\sigma)$ as the start time and completion time of job $i$ in a solution $\sigma$ (permutation-based schedule). Suppose preemption is not allowed, we have $C_i(\sigma) = S_i(\sigma) + \sum_{j=1}^{M} p_{ij}$. If there exists a due date $d_i$ for job $i$, which indicates a preferred completion time, then $L_i(\sigma) = C_i(\sigma) - d_i$ can be defined as the lateness of job $i$ and the tardiness of job $i$ can be defined as $T_i(\sigma) = max\{L_i(\sigma), 0\}$. And, the earliness for job $i$ can be defined as $E_i(\sigma) = max\{-L_i(\sigma), 0\}$. Besides, we can use the indicator function $U_i(\sigma)$ to denote whether job $i$ is tardy ($U_i(\sigma) = 1$) or not ($U_i(\sigma) = 0$). For multi-objective scheduling, assuming $w_i$ is a possible weight associated to job $i$, the following objective functions appear frequently in literature: 1) maximum completion time or makespan $C_{max}(\sigma) = max_i C_i(\sigma)$; 2) mean (weighted) completion time $\bar{C}^{(w)}(\sigma) = \frac{1}{J} \sum_{i=1}^{J} w_i C_i(\sigma)$; 3) maximum tardiness $T_{max}(\sigma) = max_i T_i(\sigma)$; 4) mean (weighted) tardiness $\bar{T}^{(w)}(\sigma) = \frac{1}{J} \sum_{i=1}^{J} w_i T_i(\sigma)$; 5) maximum earliness $E_{max}(\sigma) = max_i E_i(\sigma)$;

6) mean (weighted) earliness $\bar{E}^{(w)}(\sigma) = \frac{1}{J}\sum_{i=1}^{J} w_i E_i(\sigma)$; 7) maximum flow-time $F_{max}(\sigma) = max_i(C_i(\sigma) - S_i(\sigma))$; 8) mean (weighted) flow-time $\bar{F}^{(w)} = \frac{1}{J}\sum_{i=1}^{J} w_i(C_i(\sigma) - S_i(\sigma))$; 9) number of tardy jobs $N_T(\sigma) = \sum_{i=1}^{J} U_i(\sigma)$; and so on.

So far, many approaches have been proposed for single-objective FSSP, where exact techniques are applicable only to small-scale problems in practice and the qualities of constructive heuristics are often not satisfactory [10]. During the past decade, meta-heuristics have gained wide research for PFSS, such as Simulated Annealing (SA) [11], Genetic Algorithm (GA) [12], Evolutionary Programming (EP) [12], Tabu Search (TS) [13] and a variety of hybrid algorithms [14]. Until the late 1980s, however, it was a common practice that only one objective function was taken into account. In practice, quality may be a multi-dimensional notion. So, it is particularly important to develop effective and efficient approaches for multi-objective scheduling problems [15].

### 2.2.2 GA-based multi-objective PFSS

Currently, multi-objective optimization and scheduling based on GA is a hot topic in the fields of both Operation Research and Computer Science. Generally speaking, a multi-objective optimization problem can be formulated as follow:

$$\begin{array}{ll} Min & f(x) = (f_1(x), \ldots, f_n(x)) \\ s.t. & x \in X \end{array} \tag{2.1}$$

where $f_1, \ldots, f_n$ are the objective functions, $x$ is the vector of variables and $X$ is the set of feasible solutions.

A feasible solution $x_1$ is said to dominate another feasible solution $x_2$ (marked $x_1 \succ x_2$) if

$$\forall j \in \{1, 2, \ldots, n\}, f_j(x_1) \leq f_j(x_2) \tag{2.2}$$

and

$$\exists k \in \{1, 2, \ldots, n\}, f_j(x_1) < f_j(x_2) \tag{2.3}$$

A solution $x^*$ is Pareto optimal (non-dominated) if there is no such $x \in X$ that satisfies $x \succ x^*$.

Obviously, some objectives mentioned in Sect. 2.2.1 are conflicting, that is to say, the improvement in one objective may worsen another. Due to the multi-objective nature, scheduling approaches should efficiently and effectively find those solutions that satisfy multiple objectives. In other words, the obtained solutions should be of good proximity and diversity in sense of multi-objective optimization. Proximity means that the optimization algorithm is of excellent searching ability to obtain good solutions on or close to the optimal Pareto front. Diversity means that the algorithm is capable to obtain solutions distributed uniformly to some extent for the decision-maker to find a comparatively satisfying solution close to his preference at any time.

As we know, GA is an iterative process of "generate solutions-evaluate solutions". For multi-objective optimization problems, the generation process for solutions can be the same as that for single-objective problems, while the evaluation process may be quite different from that for single-objective problems. Basically, there are three different alternatives to handle objectives for multi-objective problems [16].

### A. Combination of objectives

This is one of the "classical" methods. A utility function or weighted function is generated to aggregate multiple objectives into a single one, thus a multi-objective flow shop problem can be converted to a single objective problem. Let $\Lambda = \{\lambda = (\lambda_1, \ldots, \lambda_n) | \lambda_j \geq 0, \sum_{j=1}^{n} \lambda_j = 1\}$ be a set of weights, denote $f_j^* = min_x\{f_j(x)\}$ as ideal point, $j = 1, \ldots, n$ and denote $\rho$ as a sufficiently small number, the following functions have been used as search directions to guide multi-objective GAs [17–23]:

Weighted Tchebycheff function:

$$s_\infty = max_{j=1,\ldots,n}\{\lambda_j(f_j(x) - f_j^*)\} \tag{2.4}$$

Augmented weighted Tchebycheff function:

$$s_{\infty aug} = max_{j=1,\ldots,n}\{\lambda_j(f_j(x) - f_j^*)\} + \rho \sum_{j=1}^{n} \lambda_j(f_j(x) - f_j^*) \tag{2.5}$$

Weighted linear sum function:

$$s_1 = \sum_{j=1}^{n} \lambda_j f_j(x) \tag{2.6}$$

For multi-objective FSSP, a class of multi-objective genetic local search (MOGLS) algorithms were proposed in [18–23], where the weight vectors were generated randomly or systematically and used to evaluate the quality of generated solutions as well as to guide the procedure of local search. In [18], a vector of weights was generated at random according to Eq. 2.7, where $rnd_j$ is a random value uniformly distributed between 0 and 1, and all the individuals in the population were evaluated according to Eq. 2.6. Then two individuals were selected with a probability to produce one offspring. A secondary population was used to store the non-dominated solutions, and some individuals of it were directly copied to the next generation. Moreover, the randomly generated weights specified different search directions towards the optimal Pareto front.

$$\lambda_j = \frac{rnd_j}{\sum_{j=1}^{n} rnd_j}, \quad j = 1, \ldots, n \tag{2.7}$$

The above algorithm was generalized by hybridizing with local search in [19], which still specified different random search directions for each selection of parents according to Eq. 2.7. After each offspring was generated using the genetic operators, a local search was applied to improve the new individual. The weight vector that was used to select the parents was still used to guide the local search of offspring. And if no parents existed (an initial generated solution), random weights were used. The local search was also implemented to some randomly selected elite solutions. Meanwhile, the maximum number of neighbors was pre-defined to control the computation time spent on the local search. It was shown that the algorithm outperformed the vector evaluated GA (VEGA) [24]. Later, Murata et al. [20,21] further extended MOGLS to a multi-objective cellular GA, where solutions were located in spatially structured cells assigned with systematically generated weights. Manhattan distance was used to measure the distance between cells and to limit the neighborhood. A new solution in a certain cell was generated by two parents, which were selected from the neighborhood. The cellular structure restricted the genetic operations to be performed on individuals not too far away. Recently, Ishibuchi et al. [22] modified their algorithm by adopting two strategies to balance local search and genetic search. First, instead of all offspring, only a few good offspring were selected to apply local search. Second, the local search direction was determined according to the location of the solution in the objective space. Besides, a local search probability was used for decreasing the number of solutions to perform local search in [23]. By examining the positive and negative effects of the hybridization with local search, the importance of striking a balance between local search and genetic search was demonstrated.

In addition, Jaszkiewicz proposed another MOGLS algorithm [17], in which every generation started by selecting random weights from a pre-specified set of weights to define a scalarizing function. Based on this function, some best solutions were selected to form a temporary population. Then two solutions were randomly selected from this temporary population to perform recombination so as to generate an offspring that was submitted to a local search. If the solution resulted by the local search was better than the worst solution in the temporary population, it was used to update the current population and the elite solution set. This algorithm was successfully applied to both several symmetric traveling salesman problems [17] and multi-objective scheduling problems [25].

### B. Alternating the objectives

This is another "classical" approach, which optimizes one objective at a time while imposing constraints on the others [16]. Usually, the order of all the objectives to be optimized should be determined, which has an effect on the success of the search. So far, this kind of method has few applications to scheduling problems.

**C. Pareto-based evaluation**

In this category, a vector containing all the objective values represents the solution fitness and the concept of Pareto dominance is used to establish preference between solutions. An algorithm was proposed by using the concept of Pareto dominance to classify the individuals in the population and a mechanism to assign suitable fitness values to promote the dispersion of the population [25]. In addition, a parallel multi-objective local search based on the concept of Pareto dominance was used to intensify the search in distinct regions. Moreover, an elite set storing all the non-dominated solutions was dynamically updated as the algorithm proceeded. The simulation results based on some flow shop scheduling problems showed that the algorithm could yield a reasonable approximation of the optimal Pareto set.

## 2.3 Quantum-inspired genetic algorithm

Recently, Han and Kim proposed some quantum-inspired genetic algorithms (QGA) [5–7], where a Q-bit representation was adopted based on the concept and principle of quantum computing. The characteristic of the representation is that any linear superposition can be represented. The smallest unit of information stored in two-state quantum computer is called a Q-bit, which may be in the "1" state, or in the "0" state, or in any superposition of the two. The state of a Q-bit can be represented as follows:

$$|\Psi\rangle = \alpha|0\rangle + \beta|1\rangle \tag{2.8}$$

where $\alpha$ and $\beta$ are complex numbers that specify the probability amplitudes of the corresponding states. $|\alpha|^2$ and $|\beta|^2$ denote the probability that the Q-bit will be found in the "1" state and "0" state respetively. Normalization of the state to the unity guarantees $|\alpha|^2 + |\beta|^2 = 1$.

A Q-bit individual as a string of $m$ Q-bit is defined as follows:

$$\left[\begin{array}{c|c|c|c} \alpha_1 & \alpha_2 & \ldots & \alpha_m \\ \beta_1 & \beta_2 & \ldots & \beta_m \end{array}\right] \tag{2.9}$$

where $|\alpha_i|^2 + |\beta_i|^2 = 1$, $i = 1, 2, \ldots, m$.

For example, for a following three-Q-bit with three pairs of amplitudes

$$\left[\begin{array}{c|c|c} 1/\sqrt{2} & 1/\sqrt{2} & 1/2 \\ 1/\sqrt{2} & -1/\sqrt{2} & \sqrt{3}/2 \end{array}\right] \tag{2.10}$$

the states can be represented as follows:

$$\frac{1}{4}|000\rangle + \frac{\sqrt{3}}{4}|001\rangle - \frac{1}{4}|010\rangle - \frac{\sqrt{3}}{4}|011\rangle + \frac{1}{4}|100\rangle + \frac{\sqrt{3}}{4}|101\rangle - \frac{1}{4}|110\rangle - \frac{\sqrt{3}}{4}|111\rangle \tag{2.11}$$

This means that the probability to represent the states $|000\rangle$, $|001\rangle$, $|010\rangle$, $|011\rangle$, $|100\rangle$, $|101\rangle$, $|110\rangle$ and $|111\rangle$ are 1/16, 3/16, 1/16, 3/16, 1/16, 3/16, 1/16, 3/16, respectively. Thus, the above three-Q-bit string contains the information of eight states. Because the Q-bit representation can represent linear superposition of state's probability, evolutionary computing with Q-bit representation has a better characteristic of population diversity than other representation.

The procedure of simple QGA proposed by Han and Kim [5–7] is described as follows.

1. Let $t=0$ and randomly generate an initial population $P_Q(t)=\{p_1^t,\dots,p_N^t\}$, where $p_j^t$ denotes the $j$-th individual in the $t$-th generation with the Q-bit representation $p_j^t = \begin{bmatrix} \alpha_1^t \big| \alpha_2^t \big| \dots \big| \alpha_m^t \\ \beta_1^t \big| \beta_2^t \big| \dots \big| \beta_m^t \end{bmatrix}$.
2. Convert $P_Q(t)$ to solutions and evaluate them, and then store the best solution $b$ among $P_Q(t)$ into the storage $B(t)$.
3. If termination condition is satisfied, then output the best solution $b$ in $B(t)$; otherwise, go on following steps.
4. Perform rotation operation, i.e. using quantum-gate $U(\theta)$ to update $P_Q(t)$ to generate $P_Q(t+1)$.
5. Convert $P_Q(t+1)$ to new solutions and evaluate them, and store the best solutions among $B(t)$ and $P_Q(t+1)$ into $B(t+1)$.
6. Store the best solution $b$ among $B(t+1)$.
7. If migration condition is satisfied, migrate $b$ (or its local one) to $B(t+1)$ globally (or locally).
8. Let $t=t+1$ and go back to *Step 3*.

Obviously, the above searching procedure is based on Q-bit representation, where the rotation operation based on quantum-gate is a main operation. In particular, a rotation gate $U(\theta)$ is employed to update a Q-bit individual as follows:

$$\begin{bmatrix} \alpha_i' \\ \beta_i' \end{bmatrix} = U(\theta_i) \begin{bmatrix} \alpha_i \\ \beta_i \end{bmatrix} = \begin{bmatrix} cos(\theta_i) & -sin(\theta_i) \\ sin(\theta_i) & cos(\theta_i) \end{bmatrix} \begin{bmatrix} \alpha_i \\ \beta_i \end{bmatrix} \tag{2.12}$$

where $(\alpha_i, \beta_i)$ is the $i$-th Q-bit, $\theta_i$ is rotation angle. $\theta_i = s(\alpha_i, \beta_i)\Delta\theta_i$, $s(\alpha_i, \beta_i)$ is the sign of $\theta_i$ that determines the direction, $\Delta\theta_i$ is the magnitude of rotation angle whose lookup table is shown in Tab. 2.1. In the Table, $b_i$ and $r_i$ are the $i$-th bits of the best solution $b$ and the binary solution $r$ respectively.

Obviously, Q-bit representation cannot be evaluated directly. For 0-1 binary optimization problems, the Q-bit representation should be converted to a binary string for evaluation. In particular, when evaluating the solution, a binary string $r$ with length $m$ is firstly constructed according to the probability amplitudes of individual $p$ with Q-bit representation, then transform the binary string $r$ to the problem solution (e.g. permutation for scheduling problems) for evaluation. In the first step, for $i = 1, 2, \dots, m$, firstly generate a random number $\eta$ between $[0, 1]$, if $\alpha_i$ of individual $p$ satisfies $|\alpha_i|^2 > \eta$, then set $r_i$ as 1, otherwise set it as 0. Based on such an evaluation way, Han

**Table 2.1.** Lookup table of rotation angle

| $r_i$ | $b_i$ | $f(r) < f(b)$ | $\Delta\theta_i$ | $s(\alpha_i, \beta_i)$ | | | |
|---|---|---|---|---|---|---|---|
| | | | | $\alpha_i\beta_i > 0$ | $\alpha_i\beta_i < 0$ | $\alpha_i = 0$ | $\beta_i = 0$ |
| 0 | 0 | flase | 0 | 0 | 0 | 0 | 0 |
| 0 | 0 | true | 0 | 0 | 0 | 0 | 0 |
| 0 | 1 | flase | 0 | 0 | 0 | 0 | 0 |
| 0 | 1 | true | $0.05\pi$ | -1 | +1 | $\pm 1$ | 0 |
| 1 | 0 | flase | $0.01\pi$ | -1 | +1 | $\pm 1$ | 0 |
| 1 | 0 | true | $0.025\pi$ | +1 | -1 | 0 | $\pm 1$ |
| 1 | 1 | flase | $0.005\pi$ | +1 | -1 | 0 | $\pm 1$ |
| 1 | 1 | true | $0.025\pi$ | +1 | -1 | 0 | $\pm 1$ |

and Kim [5–7] applied the above QGA to the knapsack problem, and Talbi et al. [26] modified the above algorithm for traveling salesman problem. For numerical optimization problems, the binary string should be further converted to a real value for evaluation. Han and Kim [7] applied a QGA for numerical optimization, and Wang et al. [27] used a QGA for parameter estimation of nonlinear systems. As we know, the solution of PFSS often should be a job permutation for evaluation. So, it may be a challenge to employ the QGA for scheduling problems, which is the main aim of this chapter.

## 2.4 Hybrid QGA for single-objective PFSS

In this Section, we first introduce the implementations of QGA and PGA for PFSS respectively, and then present the hybrid QGA.

### 2.4.1 QGA for single-objective PFSS

#### A. Representation

In QGA, a Q-bit representation $p_j^t = \begin{bmatrix} \alpha_1^t \big| \alpha_2^t \big| \ldots \big| \alpha_m^t \\ \beta_1^t \big| \beta_2^t \big| \ldots \big| \beta_m^t \end{bmatrix}$ is adopted to perform quantum-based rotation and genetic search.

#### B. Calculation of scheduling objective

Searching solution is a Q-bit string in QGA, which cannot be directly used to calculate scheduling objectives. According to the evaluation way mentioned in Sect. 2.3, a Q-bit string is firstly converted to a binary string, and then the binary string is further converted to a random key representation [28]. For example, considering a 3-job, 3-machine problem (we use 3 bits in Q-bit string to represent a job), suppose the resultant binary string is [0 1 1|1 0 1|1 0 1], thus its random key representation is [3 5 5]. Obviously, such a random key representation can be easily converted to a job permutation $\sigma$. In particular, if two random key values are different, let smaller random key denote the job with smaller number; otherwise, let the one first appears denotes the job with smaller number. Based on such a rule, the job permutation will be [1 2 3]. Once the job permutation is constructed, the scheduling objective can be calculated. Obviously, if enough Q-bits are used to represent a job, any job permutation would be constructed with the above strategy from binary representation based space.

### C. Selection

In QGA, all individuals of the population are firstly ordered from the best to the worst, then the top $N/5$ individuals are copied and the bottom $N/5$ individuals are discarded to maintain the size of population, $N$. Thus, good individuals have more chance to be reserved or to perform evolution.

### D. Crossover

In QGA, one position is randomly determined (e.g. position $i$), and then the Q-bits of the parents before position $i$ are reserved while the Q-bits after position $i$ are exchanged.

### E. Mutation

In QGA, one position is randomly determined (e.g. position $i$), and then the corresponding $\alpha_i$ and $\beta_i$ are exchanged.

### F. Catastrophe operation

To avoid premature convergence, a catastrophe operation is used in QGA: if the best solution does not change in certain consecutive generations, we regard it is trapped in local optima, then the best solution is reserved and the others will be replaced by solutions randomly generated.

### G. Rotation operation

This operation is implemented as the same way in Sect. 2.3.

### H. Procedure of QGA

The procedure of QGA for PFSS is summarized as follows.

1. Randomly generate an initial population $P_Q(t) = \{p_1^t, \ldots, p_N^t\}$, where $p_j^t$ denotes the $j$-th individual in the $t$-th generation with the Q-bit representation $p_j^t = \begin{bmatrix} \alpha_1^t \Big| \alpha_2^t \Big| \ldots \Big| \alpha_m^t \\ \beta_1^t \Big| \beta_2^t \Big| \ldots \Big| \beta_m^t \end{bmatrix}$.
2. Evaluate each solution of $P_Q(t)$, and then record the best one denoted by $b$.
3. If stopping condition is satisfied, then output the best result; otherwise go on following steps.
4. Perform selection and quantum crossover, mutation for $P_Q(t)$ to generate $P'_Q(t)$.
5. If catastrophe condition is satisfied, perform catastrophe for $P'_Q(t)$ to generate $P_Q(t+1)$ and go to *Step 7*; otherwise go to *Step 6*.
6. Applying rotation gate $U(\theta)$ to update $P'_Q(t)$ to generate $P_Q(t+1)$.
7. Evaluate every individual of $P_Q(t+1)$, and update the best solution $b$ if possible. Then let $t = t+1$ and go back to *Step 3*.

### 2.4.2 Permutation-based GA

Although quantum search has high parallel property, Q-bit representation should be converted to permutation solution for evaluation. As we know, permutation encoding is the most widely used in GA for scheduling problems due to its simplicity and easy implementation [8]. To improve the performance of QGA, permutation-based GA (PGA) is also applied as a complementary search for QGA. The implementation of PGA is briefly described as follow.

#### A. Representation

Job permutation is used as representation in PGA. For example for a 3-job flow shop problem, job permutation [2 3 1] denotes the processing sequence is job-2 first, then job-3, and job-1 last. Obviously, for such an encoding scheme, any TSP-based genetic operators can be applied.

#### B. Rank-based selection

The widely used proportional selection operator of GA is based on fitness value so that one needs to determine a transfer from objective value to fitness value. Here, the rank-based selection [12] is strongly suggested to avoid the transfer. That is, all solutions in the population are ordered from the smallest objective value to the largest one, and let $2^{N-i}/(2^N - 1)$ be the selected-probability of the $i$th solution. And the crossover operator is only performed for these selected solutions.

### C. Crossover

The crossover operator used here is partial mapping crossover (PMX), which may be the most popular one for operating the permutation [8]. In the step of PMX, firstly two crossover points are chosen and the sub-sections of the parents between the two points are exchanged, then the chromosomes are filled up by partial map. For example, for a 9-job flow shop problem, let 3 and 7 be two crossover points for parents (2 6 4 7 3 5 8 9 1) and (4 5 2 1 8 7 6 9 3). Then, the children will be (2 3 4| 1 8 7 6| 9 5) and (4 1 2| 7 3 5 8| 9 6). It has been demonstrated that PMX satisfies the fundamental properties enunciated by Holland, namely that it behaves in such a way that the best schemata reproduce themselves better than the others. Moreover, in order to generate $N$ new temporary solutions for the next mutation operator, it needs to perform such rank-based selection and crossover $N/2$ times.

### D. Mutation

Mutation serves to maintain diversity in population. In PGA, SWAP is used, i.e., two distinct elements are randomly selected and swapped.

### E. Population updating

In PGA, $(\mu + \lambda)$ strategy [3] is used as updating strategy. That is, the top $N$ solutions among the old population and the new population are selected as the population for the next generation.

### F. Procedure of PGA

The procedure of PGA is summarized as follows.

1. Let $t = 0$, initialize population $P_P(t)$, let $P_{Pt}(t) = P_P(t)$ and go to *Step 4*.
2. If the stopping criterion is satisfied, then output the best solution; otherwise go on following step.
3. Denote $P_P(t)$ as the population obtained by QGA. Then, combine $P_P(t)$ with $P''_P(t)$, and select the best half as population $P_{Pt}$.
4. Apply selection, crossover and mutation for $P_{Pt}(t)$ to generate $P'_P(t)$, and then evaluate population $P'_P(t)$.
5. Combine $P'_P(t)$ with $P_{Pt}(t)$, and select the best half as population P" (t +1).
6. Let $t = t + 1$, go to *Step 2*.

### 2.4.3 Hybrid QGA for PFSS

To achieve good results for solving PFSS, we combine the above QGA and PGA to develop a hybrid QGA (HQGA), whose framework is illustrated in Fig. 2.1.

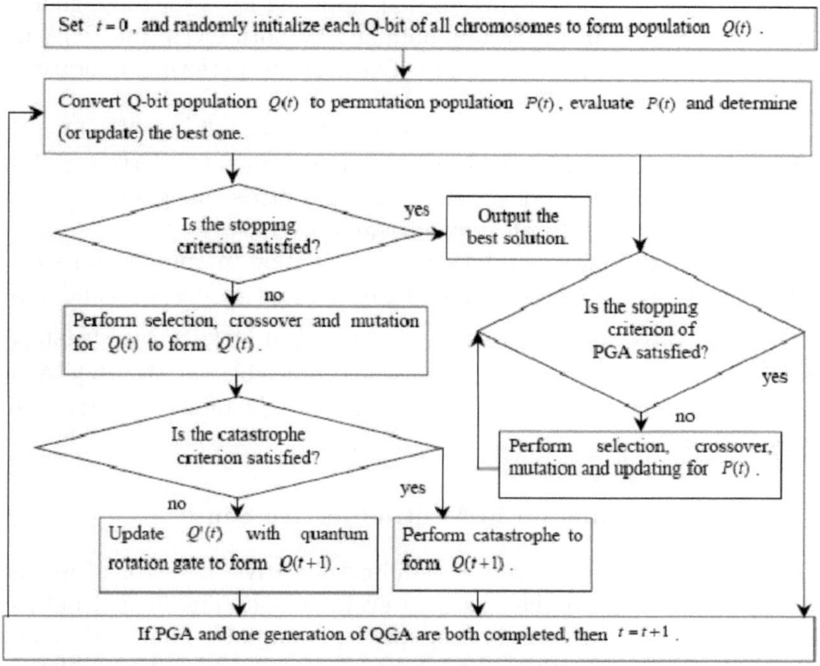


**Fig. 2.1.** The frame work of HQGA for single-objective PFSS

From Fig. 2.1, it can be seen that the main searching process is Q-bit based QGA, which sequentially performs selection, crossover, mutation and rotation operations for Q-bit based population and applies catastrophe operation to avoid being trapped in local optima. Moreover, permutation population performs classic GA as a supplement search. In brief, the HQGA can be viewed as a fusion of micro-space based search (Q-bit based search) and macro-space based search (permutation based search), thus searching behavior can be enriched, and exploration and exploitation abilities can be well balanced and enhanced. In Sect. 2.7, the simulation results about HQGA for PFSS and comparisons will be provided.

## 2.5 Hybrid QGA for multi-objective FSSP

Our idea for designing the hybrid QGA for multi-objective PFSS in this section can be summarized as follows:

1. Based on Q-bit representation, applying QGA to perform exploration in hyper 0-1 space by using quantum-based rotation and genetic operations.
2. Applying random key strategy to convert Q-bit string to job permutation for evaluation.

3. Based on permutation representation, applying PGA for those good solutions (selected from the results obtained by both QGA and PGA in the previous generation) in multi-objective sense to perform exploration and exploitation in schedule space.
4. To evaluate solutions in multi-objective sense, a randomly weighted linear sum function is used in QGA, while a non-dominated sorting technique including classification of Pareto fronts and fitness assignment is applied in PGA regarding to both proximity and diversity.
5. To maintain diversity of solutions in population, permutation-based and objective-based population trimming techniques are designed and applied.

Next, we will describe the elements of QGA and PGA for multi-objective PFSS in detail, and then present the whole procedure of the HQGA. Note that, the Representation and the Calculation of scheduling objective are the same as before.

### 2.5.1 Solution Evaluation in Multi-objective Sense

In QGA, after calculating every scheduling objective, a vector of weights is randomly generated and then used to evaluate solutions in multi-objective sense according to the weighted linear sum function (2.6). That is, the summed value is used as fitness value of a solution for multi-objective problem. The larger the value is, the better the solution is. The one with the largest summed value is regarded as the best solution (denoted $b$), which will be employed to guide the quantum-gate based rotation in current generation (see Sect. 2.3). Since the weight vector is randomly generated, it is helpful to stress search in the direction determined by the weight vector and obtain solutions distributed diversely on the Pareto front.

In PGA, the solution evaluation in multi-objective sense is realized by using non-dominated sorting technique [29–31] to stress the diversity of population and the proximity to the optimal Pareto front. The non-dominated sorting technique includes two steps, i.e., classification of Pareto fronts and fitness assignment.

In the classification step, the population is divided into $K$ sub-populations, $F_1, F_2, \dots, F_K$, according to the dominance relationship between solutions, where the solutions in set $F_i$ dominate all solutions in set $F_j$ if $i < j$, while the solutions in the same set are non-dominated to each other. That is, all solutions in the population are classified into $K$ Pareto fronts, where set $F_1$ contains all the Pareto optimal solutions in the current population. According to [31], we implement such a step (named fast non-dominated classification) as follows:

1. for each solution $x_i$ in population $P$ of size $N$, set $q_i = 0$ and $S_i = \emptyset$.
2. let $i = 1$.
3. let $j = i + 1$.

4. for solution $x_j$, if $x_i \succ x_j$ then let $S_i = S_i \bigcup \{x_j\}$ and $q_j = q_j + 1$; if $x_j \succ x_i$ then let $S_j = S_j \bigcup \{x_i\}$ and $q_i = q_i + 1$.
5. if $j = N$ then go to *Step 6*; otherwise, let $j = j + 1$ and go to *Step 4*.
6. if $i = N - 1$ then go to *Step 7*; otherwise, let $i = i + 1$ and go to *Step 3*.
7. for each solution $x_i$ in population $P$, if $q_i = 0$ then let $F_1 = F_1 \bigcup \{x_i\}$.
8. let $k = 1$.
9. if $F_k \neq \emptyset$ then set $Q = \emptyset$; otherwise, go to *Step 12*.
10. for each $x_i \in F_k$, let $q_j = q_j - 1$ for each $x_j \in S_i$. If $q_j = 0$, let $Q = Q \bigcup \{x_j\}$.
11. let $k = k + 1$, let $F_k = Q$, and go to *Step 9*.
12. let $K = k - 1$.

For a problem with $n$ objectives, the above procedure needs $N(N - 1)/2$ comparisons, and each comparison is to decide the order of two solutions on $n$ objectives. So, the overall computational complexity is $O(nN^2)$.

After the above classification, all the solutions will be assigned suitable fitness values. In our algorithm, two rules are used. One is that, to solutions in different sets, fitness values of solutions in set $F_i$ should be greater than those in set $F_j$ if $i < j$. This rule stresses the proximity, and those solutions closer to the optimal Pareto front will be assigned greater fitness values. The other is that, to solutions in the same set, solutions located in the area containing fewer solutions should be assigned greater fitness values. This rule stresses the diversity, and those solutions located in sparse area will be assigned greater fitness values. According to [25] and [31], we implement such a step (named fitness assignment) as follows:

1. let $k = 1$.
2. for each solution $x_i \in F_k$, let $d(x_i) = 0$.
3. let $j = 1$.
4. according to the $j$-th objective, order the solutions in $F_k$ based on objective values in ascending order $x_1, x_2, \ldots, x_{N_k}$, where $N_k$ denotes the number of solutions in $F_k$.
5. let $d(x_1) = L$ and $d(x_{N_k}) = L$, where $L$ is a very large number.
6. for $i = 2$ to $N_k - 1$, let $d(x_i) = d(x_i) + (f_j(x_{i+1}) - f_j(x_{i-1}))$, where $f_j(x_i)$ is the $j$-th objective value of $x_i$.
7. if $j = n$ ($n$ is the number of objective functions) then go to *Step 8*; otherwise, let $j = j + 1$ and go to *Step 4*.
8. if $k = K$ then go to *Step 9*; otherwise, let $k = k + 1$ and go to *Step 2*.
9. let $k = 1$.
10. for the set $F_k$, determine the maximum distance $d_{k,max}$ and the minimum distance $d_{k,min}$ as follows: if $N_k \leq 2$ then let $d_{k,max} = 1$ and $d_{k,min} = 1$; otherwise, let $d_{k,max} = max\{d(x_i) < L, \forall x_i \in F_k\}$ and $d_{k,min} = min\{d(x_i) \neq 0, \forall x_i \in F_k\}$.
11. for each solution $x_i \in F_k$, if $d(x_i) < L$, then let $d(x_i) = d(x_i)/d_{k,min}$; otherwise, let $d(x_i) = (d(x_i) - L) + (d_{k,max} + d_{k,min})/d_{k,min}$.

12. if $k < K$, then let $k = k + 1$ and go to *Step 10*.
13. let $MaxFitness_{K+1} = 0$ (at this time $k = K$).
14. for each solution $x_i \in F_k$, let $fitness(x_i) = MaxFitness_{k+1} + d(x_i)$.
15. let $MaxFitness_k = max\{fitness(x_i), \forall x_i \in F_k\}$.
16. if $k > 1$ then let $k = k-1$ and go to *Step 14*, otherwise stop the procedure.

Obviously, Step 1 through Step 8 is used to calculate the crowding distances; Step 9 through Step 12 is used to normalize the distances; and Step 13 through Step 16 is used to assign fitness values. In Fig. 2.2, the calculation in Step 6 is illustrated when two objective functions are considered. Fig. 2.3 through Fig. 2.5 illustrate the whole above procedure when two objectives are considered, including calculation of crowding distances, normalization of crowding distances and assignment of fitness values.

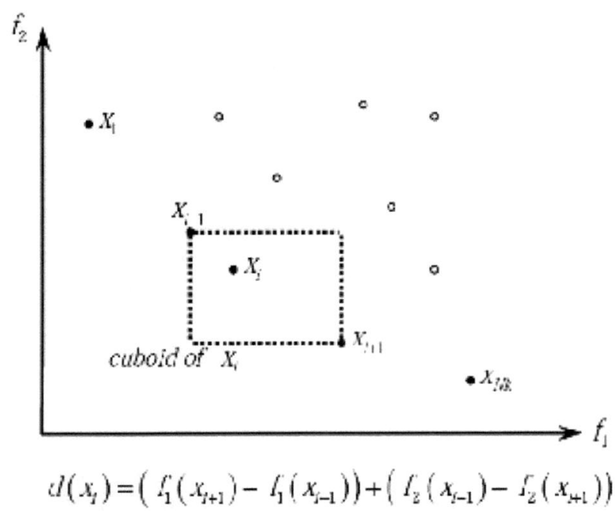


**Fig. 2.2.** Meaning of crowding distance when two objectives are considered

*Remark*: In QGA, a best solution is needed in every generation to guide the rotation operation and to determine the rotation direction and angle. The weighted linear sum function is easy and fast to determine the best solution, while the non-dominated sorting is more time-consuming. On the other hand, in PGA, fitness values are used to guide the genetic operations. Non-dominated sorting technique is more effective to assign suitable fitness values to individuals by considering both proximity and diversity. So, the weighted linear sum function and the non-dominated sorting are applied in QGA and PGA respectively to take advantage of both of their different features.

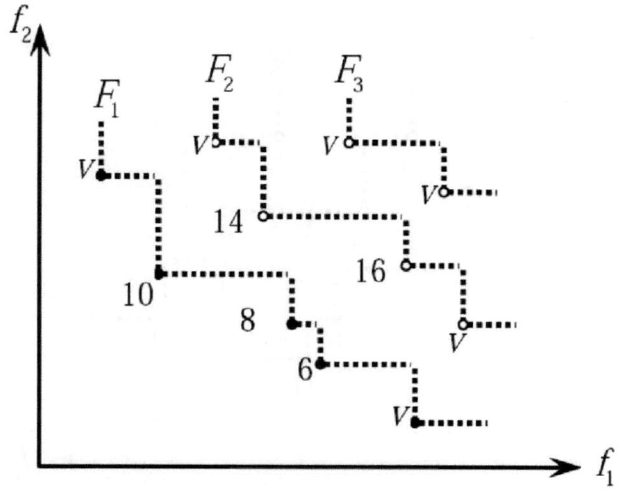


**Fig. 2.3.** Calculating crowding distances of solutions of a population

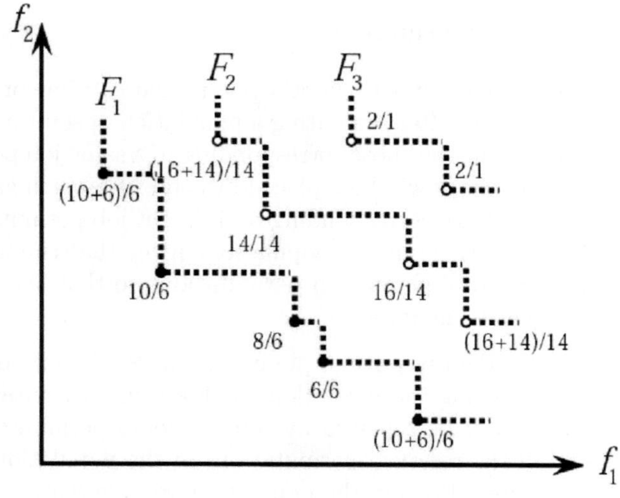


**Fig. 2.4.** Calculating normalized crowding distances of solutions of a population

### 2.5.2 Population Trimming in QGA and PGA

For a scheduling problem, evolutionary algorithms may obtain a population with some same job permutations, and the objective values (e.g. makespan) of different job permutations may be same. So, we design two trimming techniques (permutation-based and objective-based trimming techniques) to remove some redundant solutions so as to enrich the diversity of the population.

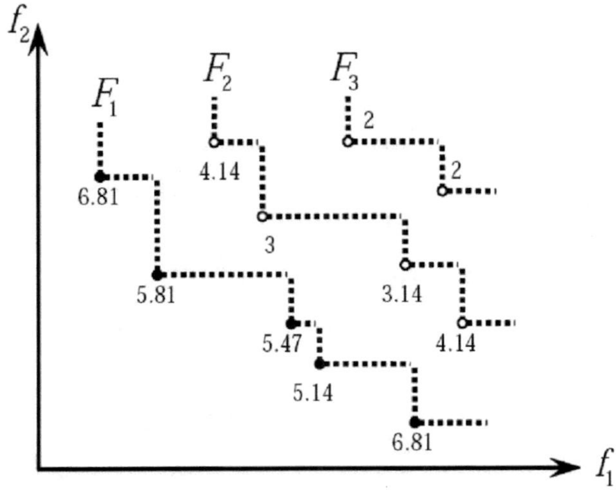


**Fig. 2.5.** Fitness assignment for solutions of a population

### A. Permutation-based Trimming

In QGA, some different Q-bit strings of a population can be converted to a same job permutation. So, after evaluating a population of solutions, only one Q-bit string is reserved among those corresponding to a same job permutation. Those removed Q-bit strings will be replaced by some Q-bit strings, which are generated randomly but are corresponding to different job permutations from all the reserved Q-bit strings in the population. After the trimming, all the strings can correspond to different job permutations so that the diversity of the population is greatly enhanced.

Similarly, to maintain the diversity of population in PGA, only one job permutation is reserved among those solutions with a same permutation. Those deleted permutations will be replaced by some random permutations, which are different from all the reserved permutations in the population. After the trimming, all the permutations in the population are different.

We denote this operator as permutation-based population trimming.

### B. Objective-based Trimming

In QGA, after evaluating a population of solutions, only one Q-bit string is reserved among those with the same objective value in every dimension. Those deleted Q-bit strings will be replaced by some random Q-bit strings, which have at least one objective value different from that of all the reserved Q-bit strings in the population. After the objective-based trimming, all the strings are different in the sense of multi-objective.

Similarly, in PGA, only one job permutation is reserved among those solutions with the same objective value in every dimension. Those deleted permutations will be replaced by some random permutations, which have at least one objective value different from all the reserved permutations in the population.

We denote this operator as objective-based population trimming.

### 2.5.3 Genetic Operators and Stopping Criteria in QGA and PGA

#### A. Selection

A simple rank-based random selection is designed in both QGA and PGA, which is different from that in Sect. 2.4. In particular, all individuals of the population are ordered from the best to the worst after evaluation in multi-objective sense. Note that the evaluation processes in QGA and PGA are different (see Sect. 2.4. Then, from the best to the worst every individual randomly chooses a partner from the remaining individuals in the population successively.

#### B. Crossover

In QGA, a two-point crossover operator is designed. In particular, two different positions are randomly determined (e.g., position $i$ and $j$). Obviously, the sub-string before the former position, the sub-string between two positions and the sub-string after the latter position can be exchanged between parents. In our algorithm, the above three ways are randomly used in hybrid sense. Since the Q-bit string is used in QGA, the above crossover can guarantee the legality of the offspring. In PGA, partial mapping crossover (PMX) is also used as Sect. 2.4.2.

#### C. Mutation

The mutation in QGA is the same as that in Sect. 2.4.1. In PGA, two different mutation operators SWAP and INSERT [8] are randomly used in hybrid sense, where SWAP randomly selects two distinct elements and swaps them, and INSERT randomly selects one element and inserts it to another different random position.

#### D. Stopping Criterion

For convenience, the maximum number of evolution generations is pre-defined in QGA and PGA respectively to determine when to stop the algorithm.

### E. Rotation Operation for Q-bit in QGA

The rotation operation is only used in QGA to adjust the probability amplitudes of each Q-bit, which is the same as that in Sect. 2.3. Note that, because $b$ is the best solution in QGA, the use of quantum-gate rotation is to emphasize the searching direction towards $b$. In addition, the weights to evaluate and determine $b$ are randomly generated, so it is helpful to enhance the diversity of searching direction.

## 2.5.4 Population Upadting in QGA and PGA

The QGA applies Q-bit representation to perform searching, while the PGA applies permutation representation to perform searching. Based on the random key method, it is easy to convert Q-bit based representation to job permutation. If copying the good solutions obtained by QGA to PGA, it is helpful to perform exploitation by PGA for these good solutions. So, we design the following way to update the populations in QGA and PGA.

In QGA, the population updated by rotation operator is used as the population for the next generation. In PGA, the individuals for the next generation are chosen from a large population, which is formed by both the current population obtained by QGA and the population of the last generation in PGA. In particular, by using the fitness assignment method mentioned above, the best half of the large population is used as the population for the next generation in PGA. Obviously, there exists an interface between QGA and PGA. That is, the good solutions obtained by QGA will have a chance to perform permutation-based search for well exploitation.

*Remark*: If copying the solutions obtained by PGA to QGA, the searching behavior of QGA may be interrupted. In addition, QGA is used to stress exploration, so it is not very necessary to copy the solutions obtained by PGA to QGA. Moreover, it is very hard to convert job permutation to suitable Q-bit string. So, the population of PGA is not copied to QGA.

## 2.5.5 Procedure of QGA and PGA

The procedures of QGA and PGA for multi-objective PFSS are described respectively as follows.

### A. Procedure of QGA

1. let $t = 0$ and randomly generate an initial population $P_Q(t) = \{p_1^t, \ldots, p_N^t\}$ where $p_j^t$ denotes the $j$-th individual (a Q-bit string) in the $t$-th generation.
2. evaluate population $P_Q(t)$, perform trimming operation for $P_Q(t)$ and record the best solution denoted by $b$.

3. if stopping condition is satisfied, then output the results; otherwise, go on following steps.
4. perform selection, crossover and mutation for $P_Q(t)$ to generate $P'_Q(t)$.
5. evaluate $P'_Q(t)$, perform trimming operation and update the best solution $b$.
6. perform rotation operation for $P'_Q(t)$ to generate $P_Q(t+1)$.
7. let $t = t + 1$ and go back to *Step 2*.

### B. Procedure of PGA

1. let $t = 0$, initialize population $P_P(t)$, let $P_{Pt}(t) = P_P(t)$ and go to *Step 4*.
2. if the stopping criterion is satisfied, then output the all the solutions in elite set $E(t)$; otherwise go on following step.
3. denote $P_P(t)$ as the population obtained by QGA. Then, combine $P_P(t)$ with $P''_P(t)$, performing trimming operation for the combined population, assign fitness values to all the individuals in the resulted population and select the best half as population $P_{Pt}(t)$.
4. apply selection, crossover and mutation for $P_{Pt}(t)$ to generate $P'_P(t)$, and then evaluate population $P'_P(t)$.
5. combine $P'_P(t)$ with $P_{Pt}(t)$, performing trimming operation for the combined population, assign fitness values to all individuals in the resulted population and select the best half as population $P''_P(t+1)$.
6. select all the non-dominated solutions in $P''_P(t+1)$ to store in $E(t)$, and delete all the dominated solutions in $E(t)$.
7. let $t = t + 1$, go to *Step 2*.

### 2.5.6 Hybrid Quantum-Inspired Genetic Algorithm

The hybrid quantum-inspired genetic algorithm for multi-objective PFSS is illustrated in Fig. 2.6.

From the framework, it can be seen that quantum-inspired search and genetic search are hybridized. On the one hand, QGA performs exploration in hyper 0-1 space by using both genetic operations and quantum rotation operation. On the other hand, PGA performs exploration in permutation-based solution space. Moreover, the good solutions obtained by QGA have chances to perform PGA for well exploitation. To maintain diversity of population, trimming techniques are applied in both QGA and PGA. To evaluate solution in multi-objective sense, the randomly weighted linear sum function is used in QGA, while fast non-dominated classification and fitness assignment techniques are used in PGA.

According to the "No Free Lunch" theorem [32], there is no method that can solve all problems optimally. So, by taking the advantages of different approaches in a hybrid way, the searching behavior can be enriched, the searching ability can be enhanced and the optimization performances can be improved.

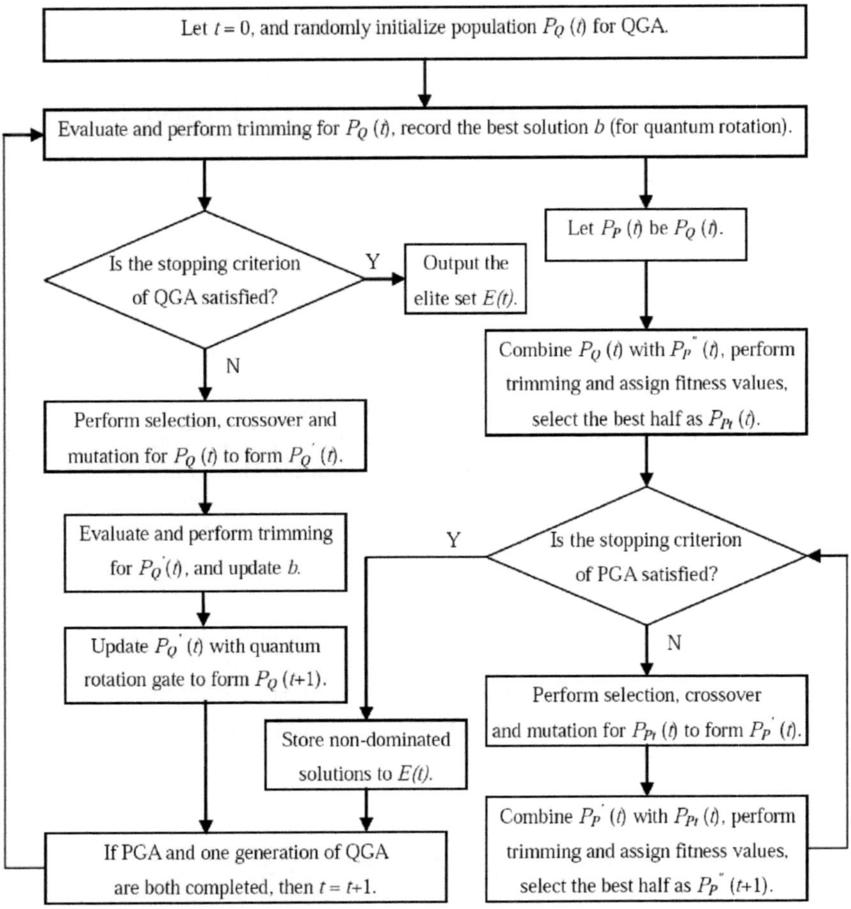


**Fig. 2.6.** The framework of HQGA for multi-objective PFSS

Due to the hybridization of two different kinds of searching algorithms, the utilization of trimming technique, and the different evaluation approaches in sense of multi-objective optimization, the HQGA will have powerful ability to obtain solutions with good proximity and diversity for multi-objective flow shop scheduling problems.

## 2.6 Performance metrics of multi-objective optimization

As for multi-objective PFSS, since the HQGA is a Pareto-based approach, the algorithm will obtain a set of non-dominated solutions $E$. Here, we introduce several metrics based on the obtained non-dominated solutions to measure the searching quality of the algorithm, which can be used for comparison between different algorithms.

### 2.6.1 Overall Non-dominated Vector Generation (*ONVG*)

For an obtained non-dominated solution set $E$, the metric *ONVG* [33] is defined as $E$, which is the number of distinct non-dominated solutions.

### 2.6.2 C Metric (*CM*)

In [34], a C metric is used to compare two non-dominated solution sets $E$ and $E'$ obtained by two different algorithms, which maps the ordered pair $(E, E')$ to the interval $[0, 1]$ to reflect the dominance relationship between solutions in the two sets:

$$C(E,E') = |\{x' \in E' | \exists : x \in E, x' \preceq x\}| / |E'| \tag{2.13}$$

Obviously, $C(E, E') = 1$ if all solutions in $E'$ are dominated by those in $E$; conversely, $C(E, E') = 0$ if all solutions in $E$ are dominated by those in $E'$. It should be noticed that the sum of $C(E, E')$ and $C(E, E')$ does not always equal to 1 because there may be some solutions in $E$ and $E'$ that are non-dominated to each other.

### 2.6.3 Distance Metrics ($D_{av}$ and $D_{max}$)

In [35] and [36], two distance metrics were used to measure the performance of non-dominated solution set $E$ relative to a reference set $R$ of the optimal Pareto front:

$$D_{av} = \sum_{x_R \in R} min_{x \in E} d(x, x_R) / |R| \tag{2.14}$$

$$D_{max} = max_{x_R \in R} \{min_{x \in E} d(x, x_R)\} \tag{2.15}$$

where $d(x, x_R) = max_{j=1,\ldots,n}\{f_j(x) - f_j(x_R)/\Delta_j\}, x \in E, x_R \in R$, and $\Delta_j$ is the range of $f_j$ values among all the solutions in set $E$ and set $R$. $D_{av}$ is the average distance from a solution $x_R \in R$ to its closest solution in $E$, and $D_{max}$ represents the maximum of the minimum distance from a solution $x_R \in R$ to any solution in $E$. Obviously, smaller $D_{av}$ and $D_{max}$ values correspond to a better approximation to the optimal Pareto front. When comparing the metric values of two non-dominated solution sets $E$ and $E'$ if the optimal Pareto front is not known, we will combine the two sets and select all the non-dominated solutions to form set $R$.

### 2.6.4 Tan's Spacing (*TS*)

In [37], the following spacing metric was used to measure how evenly the solutions are distributed:

$$TS = \sqrt{\frac{1}{|E|}\sum_{i=1}^{|E|}(D_i - \bar{D})^2/\bar{D}} \tag{2.16}$$

where $\bar{D} = \sum_{i=1}^{|E|} D_i/|E|$, $|E|$ is the number of solutions in the set $E$, $D$ is the Euclid distance in objective space between the solution $i$ and its nearest solution in the set $E$. The smaller the metric is, the more uniformly the solutions distribute.

### 2.6.5 Maximum Spread (*MS*)

This metric was defined in [37] to measure how well the true Pareto front is covered by the obtained non-dominated solutions in the set $E$ through the hyber-boxes formed by the extreme function values observed in the optimal Pareto front and $E$.

$$MS = \sqrt{\frac{1}{n}\sum_{j=1}^{n}[(max_{i=1}^{|E|} f_j(x_i) - min_{i=1}^{|E|} f_j(x_i))/(F_j^{max} - F_j^{min})]^2} \tag{2.17}$$

where $|E|$ is the number of solutions in $E$, $f_j(x_i)$ is the $j$-th objective of solution $x_i$, $F_j^{max}$, $F_j^{min}$ are the maximum and minimum values of the $j$-th objective in the optimal Pareto front. The larger the metric is, the more the optimal Pareto front is covered by solutions in the set $E$. When comparing the metric values of two sets $E$ and $E'$, if the optimal Pareto front is not known, we will combine the two sets and select all the non-dominated solutions to form optimal Pareto front.

### 2.6.6 Average Quality (*AQ*)

This metric was proposed in [38] to measure the quality of the solution set, which was originally expressed in the form of weighted Tchebycheff function $s_\infty$. But that function may hide certain aspects about the quality of solution set because poor performance with respect to proximity could be compensated by good performance in distribution of solutions. So, diversity indicators of spread and space are added to the formulation to overcome the limitation, and a metric is given by the average value of a scalarizing function over a representative sample of weight vectors as follow:

$$AQ = \sum_{\lambda \in \Lambda} s_a(f, z^0, \lambda, \rho)/|\Lambda| \tag{2.18}$$

where $s_a(f, z^0, \lambda, \rho) = min_i\{max_j\{\lambda_j(f_j(x_i) - z_j^0)\} + \rho\sum_{j=1}^{n}\lambda_j(f_j(x_i) - z_j^0)\}$ and $\Lambda = \{\lambda = (\lambda_1, \ldots, \lambda_n)|\lambda_j \in \{0, 1/r, 2/r, \ldots, 1\}, \sum_{j=1}^{n}\lambda_j = 1\}$. $z^0$ is a reference point in the objective space which is set to (0,0), and $\rho$ is a sufficiently small number which is set to 0.01 in this paper. Besides, $r$ is a parameter changed as the number of objectives, which is often set as 50 for 2-objective problems. A smaller metric corresponds to a better solution set.

In addition, to reflect the efficiency of a multi-objective optimization algorithm, running time ($t$) is also used as a metric.

The above metrics are used to measure the performances of a multi-objective optimization algorithm from different aspects. *ONVG* metric reflects the number of non-dominated solutions. $C$ metric shows the dominance relationship between two compared solution sets. Distance metrics reflect the closeness of obtained solution set to the optimal Pareto front. *TS* metric evaluates the distribution of solutions in the set. *MS* metric measures how well the solution set covers the optimal Pareto front. *AQ* metric evaluates comprehensive quality of all the solutions. Running time reflects the efficiency of the algorithm. Based on these metrics, we will evaluate the performances of the proposed HQGA and carry out some comparisons.

## 2.7 Simulation results and comparisons on single-objective PFFS

To test the performance of the proposed HQGA for single-objective PFSS, computational simulation is carried out with some well-studied benchmarks. In this paper, 29 problems that were contributed to the OR-Library are selected. The first eight problems were called car1, car2 through car8 by Carlier [39]. The other 21 problems were called rec01, rec03 through rec41 by Reeves [40], who used them to compare the performances of simulated annealing, genetic algorithm and neighborhood search and found these problems to be particularly difficult. Thus far these problems have been used as benchmarks for study with different methods by many researchers.

We set population size as 40, maximum generation (stopping condition of QGA) as $J \times M$, the length of each chromosome as $10 \times J$ (i.e., every 10 Q-bits correspond to a job), crossover probability as 1, mutation probability as 0.05, catastrophe happens in QGA if the best solution does not change in consecutive $J \times M/10$ generations.

Besides, PGA is performed 10 generations (stopping condition of PGA) in every generation of QGA. For fair comparison, pure PGA and pure QGA use the same computational effort as HQGA.

We run each algorithm 20 times for every problem, and the statistical results are summarized in Tab. 2.2, where BRE, ARE and WRE denote the best, average and worst percentage relative errors with $C^*$ (lower bound or optimal makespan) respectively.

**Table 2.2.** The statistical results of testing algorithms

| P | $J, M$ | C* | HQGA | | | NEH | PGA | | QGA | |
|---|---|---|---|---|---|---|---|---|---|---|
| | | | BRE | ARE | WRE | ARE | BER | ARE | BRE | ARE |
| Car1 | 11,5 | 7038 | 0 | 0 | 0 | 0 | 0 | 0 | 0 | 0 |
| Car2 | 13,4 | 7166 | 0 | 0 | 0 | 2.93 | 0 | 0.21 | 0 | 1.90 |
| Car3 | 12,5 | 7312 | 0 | 0 | 0 | 1.19 | 0 | 0.86 | 1.19 | 1.65 |
| Car4 | 14,4 | 8003 | 0 | 0 | 0 | 0 | 0 | 0.08 | 0 | 0.06 |
| Car5 | 10,6 | 7720 | 0 | 0 | 0 | 1.49 | 0 | 0.09 | 0 | 0.11 |
| Car6 | 8,9 | 8505 | 0 | 0 | 0 | 3.15 | 0 | 0.26 | 0 | 0.19 |
| Car7 | 7,7 | 6590 | 0 | 0 | 0 | 0 | 0 | 0 | 0 | 0 |
| Car8 | 8,8 | 8366 | 0 | 0 | 0 | 2.37 | 0 | 0.18 | 0 | 0.03 |
| Rec01 | 20,5 | 1247 | 0 | 0.14 | 0.16 | 4.49 | 1.28 | 4.45 | 2.81 | 6.79 |
| Rec03 | 20,5 | 1109 | 0 | 0.17 | 0.72 | 2.07 | 0.72 | 2.18 | 0.45 | 3.87 |
| Rec05 | 20,5 | 1242 | 0.24 | 0.34 | 2.17 | 3.14 | 0.24 | 1.50 | 2.25 | 3.00 |
| Rec07 | 20,10 | 1566 | 0 | 1.02 | 2.11 | 3.83 | 1.85 | 2.98 | 1.15 | 4.67 |
| Rec09 | 20,10 | 1537 | 0 | 0.64 | 2.41 | 2.99 | 2.60 | 4.12 | 4.03 | 6.40 |
| Rec11 | 20,10 | 1431 | 0 | 0.67 | 2.66 | 8.32 | 4.05 | 6.17 | 6.08 | 8.79 |
| Rec13 | 20,15 | 1930 | 0.16 | 1.07 | 3.06 | 3.73 | 3.01 | 4.66 | 5.08 | 7.98 |
| Rec15 | 20,15 | 1950 | 0.05 | 0.97 | 1.64 | 3.23 | 2.21 | 3.96 | 3.49 | 5.93 |
| Rec17 | 20,15 | 1902 | 0.63 | 1.68 | 2.73 | 6.15 | 3.26 | 5.86 | 6.51 | 9.10 |
| Rec19 | 30,10 | 2093 | 0.29 | 1.43 | 2.58 | 4.40 | 5.88 | 7.39 | 7.98 | 9.80 |
| Rec21 | 30,10 | 2017 | 1.44 | 1.63 | 1.83 | 5.65 | 5.16 | 7.05 | 6.94 | 10.05 |
| Rec23 | 30,10 | 2011 | 0.50 | 1.20 | 2.64 | 7.16 | 5.92 | 7.56 | 9.10 | 10.55 |
| Rec25 | 30,15 | 2513 | 0.77 | 1.87 | 3.34 | 5.21 | 6.29 | 7.85 | 7.16 | 10.06 |
| Rec27 | 30,15 | 2373 | 0.97 | 1.83 | 2.82 | 5.27 | 6.41 | 8.14 | 7.63 | 11.05 |
| Rec29 | 30,15 | 2287 | 0.35 | 1.97 | 3.63 | 4.55 | 8.53 | 10.53 | 12.42 | 14.06 |
| Rec31 | 50,10 | 3045 | 1.05 | 2.50 | 4.07 | 4.20 | 8.83 | 10.16 | 9.82 | 12.68 |
| Rec33 | 50,10 | 3114 | 0.83 | 0.91 | 1.19 | 4.08 | 5.84 | 7.07 | 6.20 | 9.54 |
| Rec35 | 50,10 | 3277 | 0 | 0.15 | 0.34 | 1.10 | 2.72 | 3.88 | 4.21 | 6.52 |
| Rec37 | 75,20 | 4951 | 2.52 | 4.33 | 5.66 | 5.58 | 13.78 | 15.27 | 15.54 | 17.49 |
| Rec39 | 75,20 | 5087 | 1.65 | 2.71 | 4.46 | 4.34 | 12.04 | 13.06 | 13.50 | 15.49 |
| Rec41 | 75,20 | 4960 | 3.13 | 4.15 | 5.73 | 6.69 | 13.48 | 15.86 | 16.92 | 18.84 |

From Tab. 2.2, it can be seen that the results obtained by HQGA are much better than not only that of NEH heuristic but also that of PGA and QGA. Even the AVE values resulted by HQGA are better than BRE values resulted by PGA and QGA.

Secondly, the BRE values resulted by HQGA are very near to 0, especially for small-scale problems, which means HQGA is able to obtain good solutions in global sense.

Thirdly, the BRE, ARE and WRE values resulted by HQGA are closer than those by other methods, which means HQGA has good robustness and consistence on initial conditions.

So, it is concluded that HQGA is an effective approach for single-objective PFSSs.

## 2.8 Simulation results and comparisons on multi-objective PFFS

In this Section, not only some multi-objective testing problems in the literature but also some randomly generated testing problems are used. As for the former cases, we refer to [41–46] for detail. As for the latter cases, there exists a classical way to randomly generate instances of scheduling problems [25,47–49]. In particular, the processing time of each job in every machine is uniformly distributed in interval $[1, 99]$. And the due date of each job is uniformly distributed in interval $[Q(1-u-v/2), Q(1-u+v/2)]$, where $u$ and $v$ respectively represent the tardiness factor of jobs and dispersion range of due dates, and $Q$ is a following lower bound of makespan [50]:

$$Q = max\{max_{1\le j\le M}\{\sum_{i=1}^{J} p_{ij} + min_i \sum_{l=1}^{j-1} p_{il} + min_i \sum_{l=j+1}^{M} p_{il}\}, max_i \sum_{j=1}^{M} p_{ij}\} \tag{2.19}$$

In this paper, four scenarios about due date are considered [25], where each scenario is determined by a different combination of the values of $u$ and $v$ . It is said that when $u$ increases the due dates are more restrictive, while when $v$ increases the due dates are more diversified.

1. low tardiness factor ($u = 0.2$) and small due date range ($v = 0.6$);
2. low tardiness factor ($u = 0.2$) and wide due date range ($v = 1.2$);
3. high tardiness factor ($u = 0.4$) and small due date range ($v = 0.6$);
4. high tardiness factor ($u = 0.4$) and wide due date range ($v = 1.2$).

### 2.8.1 Testing on existing problems in the literature

In this part, based on some testing problems with different scales and different types of objectives from [41–46], we test the performance of the HQGA on PC with P4-2.40GHz, 256M-RAM, WIN-XP OS and VC++ software to demonstrate its generality and applicability to multi-objective scheduling problems. For these testing problems, the HQGA is designed as follows. In QGA, population size is 10, the maximum generation is 10, crossover probability is 1, mutation probability is 0.05, the length of Q-bit string is $5 \times J$ (i.e., every 5 Q-bits correspond to a job). In PGA, population size is 10, which is the same as that of QGA, crossover and mutation probabilities both are 1, and only one generation of PGA is performed in every generation of HQGA (that is, the maximum generation of PGA is 1). Since these problems are easy to solve, we do not apply the trimming technique in HQGA.

In Tab. 2.3, the results of nine problems by both HQGA and the methods in the literatures are presented. From Tab. 2.3, it can be obviously found that HQGA can obtain identical or even better non-dominated solutions. So, the proposed HQGA is an effective approach for multi-objective scheduling.

**Table 2.3.** Results on testing problems found from literature

| Problems | Author's Methods | Objective | Objective values of obtained solutions | |
|---|---|---|---|---|
| | | | By author | By HQGA |
| Nelson et al. [41] $j=6, M=1$ | Exact | $(\bar{F}, N_T)$ | (130, 1) (103, 2) (98, 3) (93, 4) | Idem author |
| | | $(N_T, T_{max})$ | (3, 50)(2, 95) (1, 185) | Idem author |
| | | $(\bar{F}, T_{max})$ | (93, 55) (129, 50) | Idem author |
| Shanthikumar [42], $J=8, M=1$ | Exact | $(T_{max}, N_T)$ | (44,3) | (44, 3) (36,4) (29, 5) |
| Selen and Hott [43], $J=6, M=4$ | Exact | $(C_{max}, \bar{F})$ | (78,53) | Idem author |
| Ho and Chang [44], $J=5, M=4$ | Heuristic | $(C_{max}, \bar{F})$ | (213, 56) | (213, 56) (234, 153) |
| Rajendran [45], $J=5, M=3$ | Heuristic | $(C_{max}, \bar{F})$ | (52,29) (51, 31) | (51, 29.4) |
| Liao et al. [46][a], $J=6, M=2$ | Exact | $(C_{max}, N_T)$ | (29,1) | Idem author |
| | | $(C_{max}, \bar{T})$ | (29,2) | Idem author |

*Remark*: Liao et al. [46] gave the lower bound for $C_{max}, N_T$ and $\bar{T}$, given the partial sequence [1, 4] always at the end of the sequences. Here, HQGA give the same results by scheduling the four remaining tasks and adding tasks 1 and 4 at the end of the solutions obtained.

### 2.8.2 Testing and comparisons on randomly generated instances

In this part, we test the performances of the HQGA based on some randomly generated 2-objective instances with 20-job and 10-machine. For each of the four scenarios for the problem, 5 random instances are used. The HQGA is designed as follows. In QGA, population size is 100, the maximum generation is $J \times M$ (200), crossover probability is 1, mutation probability is 0.05, length of Q-bit string is $5 \times J$ (i.e., every 5 Q-bits correspond to a job). In PGA, population size is 100, which is the same as that of QGA, crossover and mutation probabilities both are 1, and only one generation of PGA is performed in every generation of HQGA (that is, the maximum generation of PGA is 1). To show the effectiveness of hybridization, a pure PGA by removing all the elements of QGA in HQGA and a pure QGA by removing all the elements of PGA in HQGA, are used for comparison, where all parameters and operators are the same as those in HQGA. In order to fairly compare the algorithms, the maximum generations of pure PGA and pure QGA are set to 400, which equals the total generation of PGA and QGA in HQGA.

Firstly, consider an instance under scenario 3 with two objectives (makespan and average completion time). And permutation-based trimming technique is used in HQGA, pure PGA and pure QGA. In Fig. 2.7, typical running results by the three algorithms are illustrated; and in Fig. 2.8, the results by HQGA and pure PGA are shown more clearly. Obviously, HQGA can obtain Pareto solutions with better proximity than both pure PGA and pure QGA, and most solutions obtained by HQGA can dominate the solutions obtained by both pure PGA and pure QGA. Moreover, it can be seen that HQGA can obtain more Pareto solutions, and these solutions obtained by HQGA are of good diversity. Since the performance of pure QGA is not very satisfactory, which is even worse than that of pure PGA, we do not provide the results of pure QGA in the following simulations but only compare the results of HQGA and pure PGA.

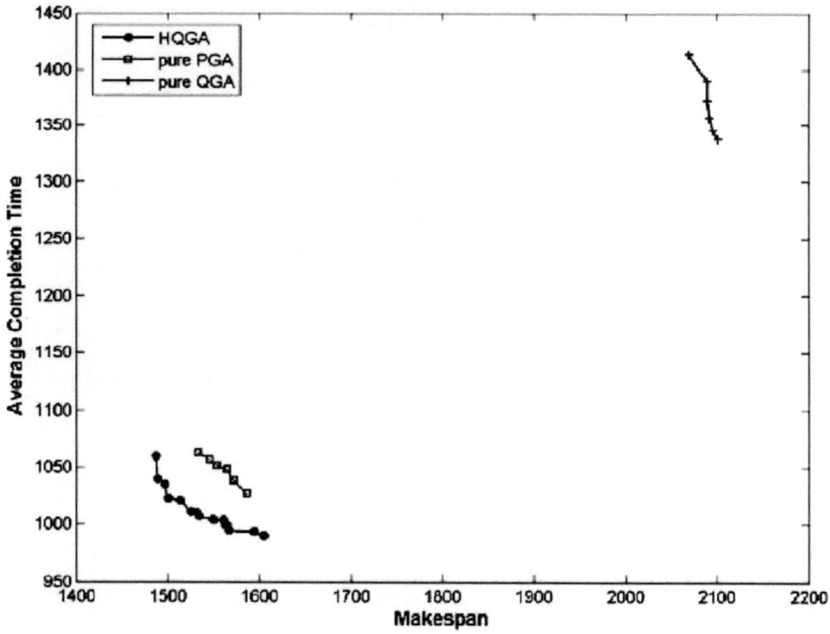


**Fig. 2.7.** Pareto fronts produced by HQGA, pure PGA and pure QGA

In Fig. 2.9 and Fig. 2.10, the changes of the non-dominated set produced by HQGA and pure PGA during the evolution are illustrated respectively, from which the changes of the Pareto front can be seen clearly. Moreover, Fig. 2.11 and Fig. 2.12 illustrate the changes of the number of non-dominated solutions produced by HQGA and pure PGA as evolution proceeds. Obviously, the number of non-dominated solutions produced by HQGA not only is larger but also changes more frequently than that by pure PGA. So, it is concluded that HQGA is of better ability than pure PGA to obtain solutions with good proximity and diversity in sense of multi-objective optimization.

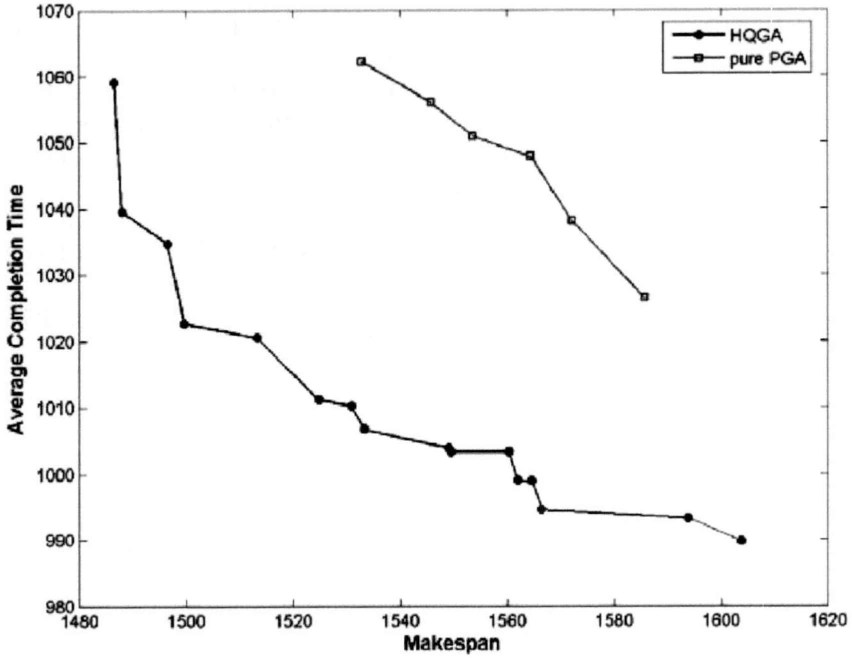


**Fig. 2.8.** Pareto fronts produced by HQGA and pure PGA

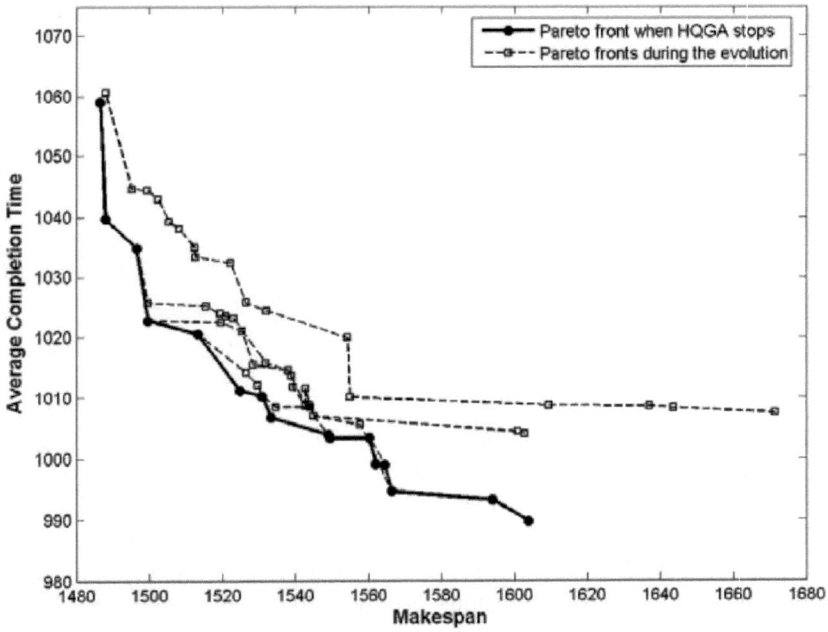


**Fig. 2.9.** Pareto fronts produced by HQGA during the evolution

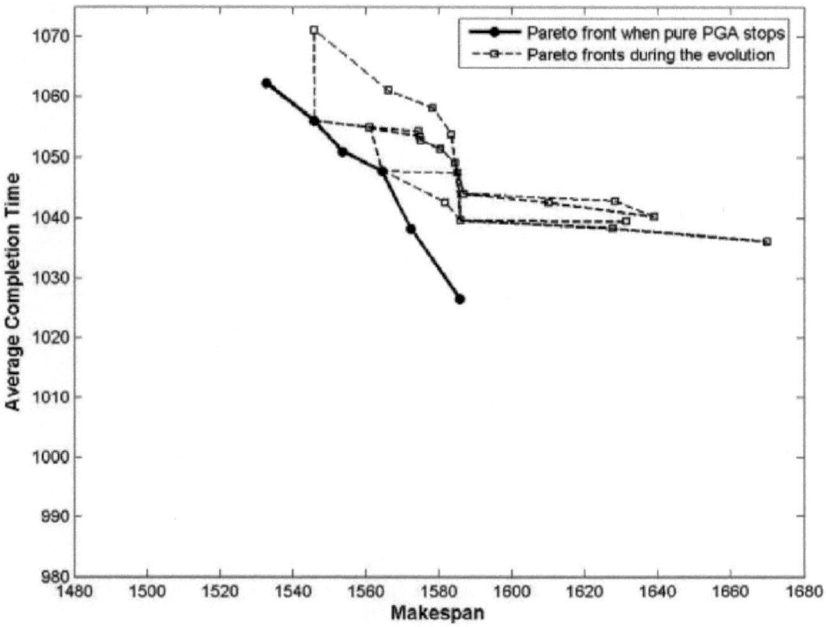


**Fig. 2.10.** Pareto fronts produced by pure PGA during the evolution

Still based on the above problem (scenario 3), we compare the performance metrics of the solution set obtained by HQGA and pure PGA. Tab. 2.4 illustrates the values of all performance metrics.

**Table 2.4.** Metric comparison for Scenario 3 when considering $f = (C_{max}, \bar{C})$ (permutation-based trimming)

| Metric | Method | Exp.1 | Exp.2 | Exp.3 | Exp.4 | Exp.5 | Average |
|---|---|---|---|---|---|---|---|
| $ONVG$ | HQGA | 18 | 16 | 14 | 12 | 25 | 17 |
| | pure PGA | 10 | 6 | 4 | 11 | 15 | 9 |
| $CM$ | HQGA | 0 | 0 | 0 | 0 | 0 | 0 |
| | pure PGA | 1 | 1 | 1 | 1 | 1 | 1 |
| $D_{av}$ | HQGA | 0.0000 | 0.0000 | 0.0000 | 0.0000 | 0.0000 | 0.0000 |
| | pure PGA | 0.3514 | 0.3977 | 0.4399 | 0.4917 | 0.3268 | 0.4015 |
| $D_{max}$ | HQGA | 0.0000 | 0.0000 | 0.0000 | 0.0000 | 0.0000 | 0.0000 |
| | pure PGA | 0.4268 | 0.5058 | 0.6738 | 0.5391 | 0.3711 | 0.5033 |
| $TS$ | HQGA | 0.8010 | 0.6974 | 0.9131 | 1.0768 | 1.6561 | 1.0289 |
| | pure PGA | 0.6949 | 0.2436 | 0.1263 | 0.8384 | 0.7522 | 0.5311 |
| $MS$ | HQGA | 0.6214 | 0.9788 | 0.6417 | 0.5060 | 0.8044 | 0.7105 |
| | pure PGA | 0.6657 | 0.4735 | 0.6009 | 0.7310 | 0.7357 | 0.6414 |
| $AQ$ | HQGA | 999.1268 | 959.8224 | 927.4387 | 1042.3823 | 984.7138 | 982.6968 |
| | pure PGA | 1033.2858 | 991.4405 | 954.4940 | 1068.6533 | 1011.4361 | 1011.8619 |
| $t$ | HQGA | 43.7810 | 44.5460 | 43.6400 | 42.8280 | 44.1710 | 43.7932 |
| | pure PGA | 36.3900 | 35.6090 | 35.7650 | 35.9370 | 36.7180 | 36.0838 |

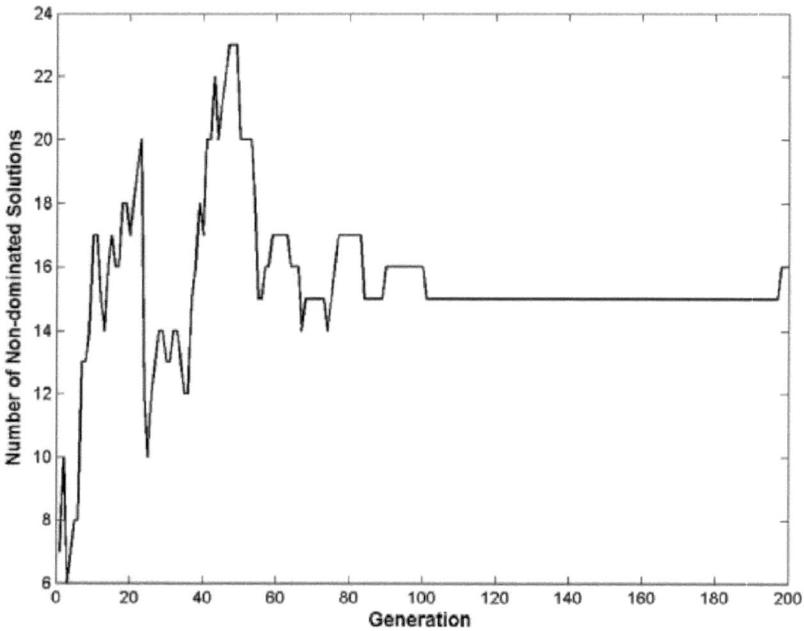


**Fig. 2.11.** Number of non-dominated solutions produced by HQGA as evolution proceeds

First, from Tab. 2.4 it can be seen that the $ONVS$ values of HQGA are always larger than those of pure PGA, which means the non-dominated solutions obtained by HQGA is always more than that of pure PGA.

Second, it can be seen that the $CM$ values of HQGA are 0 while those of pure PGA are 1 in all the cases, so it can be concluded that the solutions obtained by HQGA can dominate those obtained by pure PGA. Considering the $ONVS$ and $CM$ metrics, it can be concluded that HQGA can find a larger number of non-dominated solutions with better proximity than pure PGA.

Third, it can also be seen that $D_{av}$ and $D_{max}$ values of HQGA are always 0, while these two metric values of pure PGA are very large. Since $D_{av}$ and $D_{max}$ metrics measure the distance between the obtained solution set and the optimal Pareto front in average and max-min senses respectively, so it can be concluded that the solutions obtained by HQGA are closer to the optimal Pareto front than those obtained by pure PGA.

Fourth, it can be seen that, the $TS$ values of HQGA are comparable to or occasionally larger than those of pure PGA, and the MS values of HQGA are sometimes smaller than those of pure PGA. It seems to indicate that for some cases the solutions obtained by pure PGA have a more uniform distribution than those of HQGA, and the solution set obtained by HQGA covers less

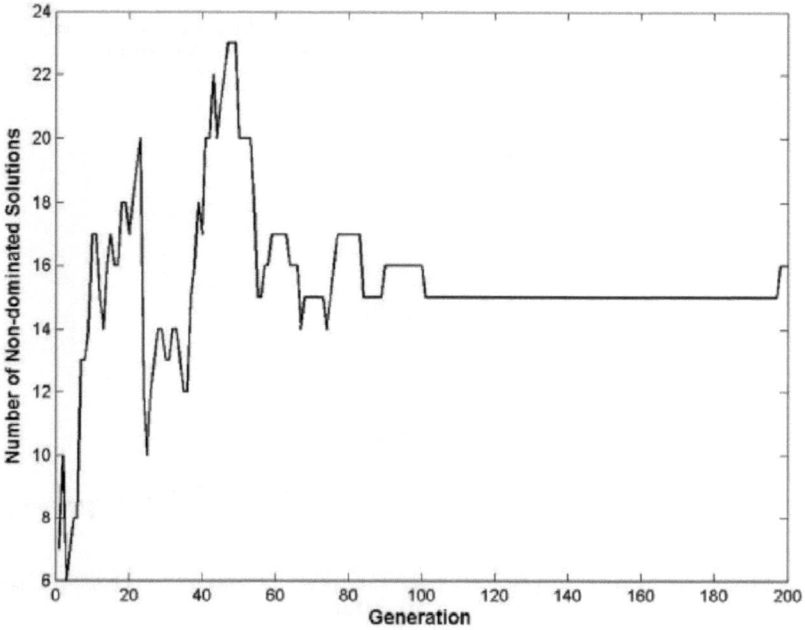


**Fig. 2.12.** Number of non-dominated solutions produced by pure PGA as evolution proceeds

area of the optimal true Pareto front than that by pure PGA. However, it should be noted that HQGA obtains more non-dominated solutions closer to the optimal Pareto front than pure PGA, any of which can influence the whole diversity performance to some extent. So, the values of $TS$ and $MS$ metrics cannot comprehensively reflect the performances of a multi-objective algorithm. Thus, we use $AQ$ metric that considers both the proximity and diversity for comparison. From Tab. 2.4, it can be seen that the $AQ$ values of HQGA are smaller than those of pure PGA for all the cases. So, it is concluded that HQGA is a more effective optimization algorithm than pure PGA.

Finally, comparing the efficiency ($t$ value), it can be seen that the running times of HQGA are only slightly larger than that of pure PGA.

Next, we carry out some simulations to show the effectiveness of trimming techniques. In Fig. 2.13, the solution sets produced by HQGA with different trimming techniques and without trimming are presented.

From Fig. 2.13, it can be seen that the solution set obtained by HQGA with either permutation-based trimming or objective-based trimming is better than that obtained by HQGA without trimming. On the one hand, all the solutions obtained by HQGA without trimming are dominated by solutions obtained by HQGA with trimming. On the other hand, the set obtained by HQGA

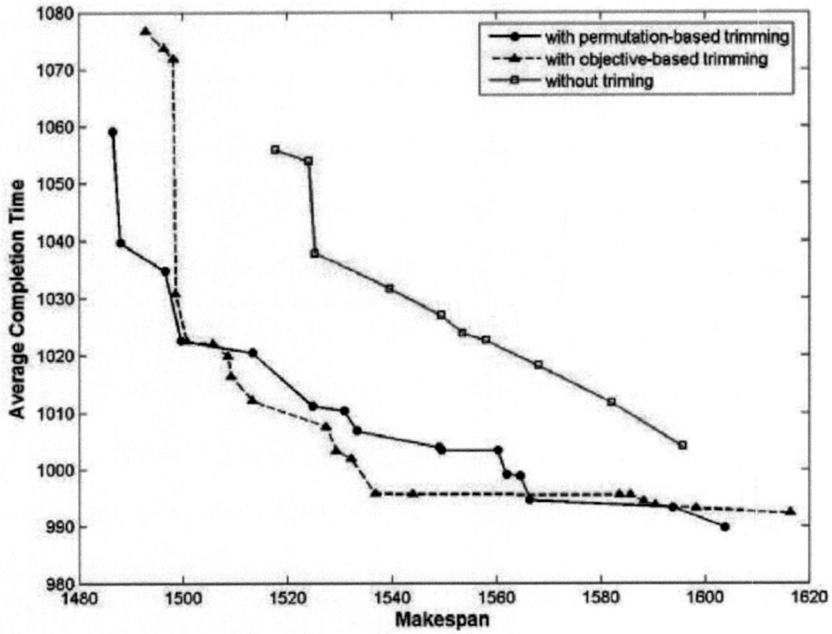


**Fig. 2.13.** Pareto fronts produced by HQGAs with permutation-based trimmings, objective-based trimming and without trimming

with trimming contains more solutions than that obtained by HQGA without trimming. So, it is concluded that trimming techniques by removing redundant solutions are helpful for HQGA to achieve good performances.

As for the two different kinds of trimming techniques, it can be found from Fig. 2.13 that there is no absolute dominance relationship between the resultant Pareto fronts. In the following simulations, we will only apply objective-based trimming in HQGA.

As we know, under different scenarios (with different $u$ and $v$ ) the magnitude and diversity of due date may be different, which will result in problems with different degrees of difficulty. So, we test the robustness of HQGA for the 20-job and 10-machine flow shop scheduling problems with the objectives of makespan and maximum tardiness under four different scenarios.

From the results in Tab. 2.5 through Tab. 2.8, it can be found that all performance metrics of HQGA are similar for different scenarios, which means HQGA can obtain solution sets with good proximity and diversity in sense of multi-objective optimization for scheduling problems with different degrees of difficulty. That is, HQGA is of good robustness for different multi-objective scheduling problems. In addition, from these tables, once again it can be seen that the performances of HQGA are better than those of pure PGA.

**Table 2.5.** Metric comparison for Scenario 1 when considering $f = (C_{max}, T_{max})$ (objective-based trimming)

| Metric | Method | Exp.1 | Exp.2 | Exp.3 | Exp.4 | Exp.5 | Average |
|---|---|---|---|---|---|---|---|
| $ONVG$ | HQGA | 17 | 15 | 19 | 19 | 25 | 19 |
| | pure PGA | 13 | 11 | 14 | 13 | 7 | 12 |
| $CM$ | HQGA | 0 | 0 | 0 | 0 | 0 | 0 |
| | pure PGA | 1 | 1 | 1 | 1 | 1 | 1 |
| $D_{av}$ | HQGA | 0.0000 | 0.0000 | 0.0000 | 0.0000 | 0.0000 | 0.0000 |
| | pure PGA | 0.3332 | 0.2032 | 0.3152 | 0.2551 | 0.1711 | 0.2556 |
| $D_{max}$ | HQGA | 0.0000 | 0.0000 | 0.0000 | 0.0000 | 0.0000 | 0.0000 |
| | pure PGA | 0.4968 | 0.2641 | 0.3812 | 0.2929 | 0.2962 | 0.3463 |
| $TS$ | HQGA | 2.3372 | 2.6377 | 0.6741 | 0.7588 | 1.4365 | 1.5689 |
| | pure PGA | 0.9239 | 1.5760 | 0.4769 | 0.7182 | 0.9089 | 0.9208 |
| $MS$ | HQGA | 0.8458 | 0.9040 | 0.5106 | 0.4372 | 0.7091 | 0.6814 |
| | pure PGA | 0.6189 | 0.8153 | 0.8462 | 0.8039 | 0.7828 | 0.7734 |
| $AQ$ | HQGA | 741.9177 | 812.4140 | 780.8477 | 818.6470 | 798.8750 | 790.5403 |
| | pure PGA | 795.8478 | 840.4334 | 810.3190 | 853.3200 | 841.8020 | 828.3445 |
| $t$ | HQGA | 27.2960 | 27.4840 | 27.9530 | 27.7500 | 27.6400 | 27.6246 |
| | pure PGA | 22.3900 | 23.3120 | 22.6250 | 23.1250 | 22.5150 | 22.7934 |

**Table 2.6.** Metric comparison for Scenario 2 when considering $f = (C_{max}, T_{max})$ (objective-based trimming)

| Metric | Method | Exp.1 | Exp.2 | Exp.3 | Exp.4 | Exp.5 | Average |
|---|---|---|---|---|---|---|---|
| $ONVG$ | HQGA | 20 | 11 | 18 | 16 | 20 | 17 |
| | pure PGA | 15 | 11 | 13 | 10 | 11 | 12 |
| $CM$ | HQGA | 0 | 0 | 0 | 0 | 0 | 0 |
| | pure PGA | 1 | 1 | 1 | 1 | 1 | 1 |
| $D_{av}$ | HQGA | 0.0000 | 0.0000 | 0.0000 | 0.0000 | 0.0000 | 0.0000 |
| | pure PGA | 0.2509 | 0.1928 | 0.1547 | 0.1593 | 0.3203 | 0.2156 |
| $D_{max}$ | HQGA | 0.0000 | 0.0000 | 0.0000 | 0.0000 | 0.0000 | 0.0000 |
| | pure PGA | 0.3559 | 0.2410 | 0.2041 | 0.2934 | 0.4621 | 0.3113 |
| $TS$ | HQGA | 0.8031 | 2.6371 | 2.0813 | 0.8262 | 2.3893 | 1.7474 |
| | pure PGA | 0.8259 | 0.7041 | 0.8289 | 0.9443 | 0.3838 | 0.7374 |
| $MS$ | HQGA | 0.6708 | 0.7235 | 0.7262 | 0.7583 | 0.9211 | 0.7600 |
| | pure PGA | 0.7777 | 0.9292 | 0.8577 | 0.8108 | 0.8862 | 0.8523 |
| $AQ$ | HQGA | 807.5711 | 837.7021 | 875.5760 | 863.2273 | 859.9027 | 848.7958 |
| | pure PGA | 854.9855 | 852.2990 | 916.4766 | 869.9304 | 890.8465 | 876.9076 |
| $t$ | HQGA | 27.7190 | 27.9060 | 28.2500 | 27.9060 | 27.9718 | |
| | pure PGA | 22.9370 | 22.6710 | 22.6710 | 23.3750 | 22.1720 | 22.7652 |

In a word, the proposed hybrid quantum-inspired genetic algorithm is also an effective approach for multi-objective PFSSs, which not only can obtain solutions with good proximity and diversity in the sense of multi-objective optimization but also is of good robustness for multi-objective scheduling problems with different degrees of difficulty.

## 2.9 Conclusions and future work

This chapter presented hybrid quantum-inspired genetic algorithms for both single-objective and multi-objective permutation flow shop scheduling. Not only was the quantum-inspired GA applied based on Q-bit representation for

**Table 2.7.** Metric comparison for Scenario 3 when considering $f = (C_{max}, T_{max})$ (objective-based trimming)

| Metric | Method | Exp.1 | Exp.2 | Exp.3 | Exp.4 | Exp.5 | Average |
|---|---|---|---|---|---|---|---|
| $ONVG$ | HQGA | 8 | 20 | 13 | 22 | 16 | 16 |
| | pure PGA | 7 | 16 | 8 | 14 | 20 | 13 |
| $CM$ | HQGA | 0 | 0 | 0 | 0 | 0 | 0 |
| | pure PGA | 1 | 1 | 1 | 1 | 1 | 1 |
| $D_{av}$ | HQGA | 0.0000 | 0.0000 | 0.0000 | 0.0000 | 0.0000 | 0.0000 |
| | pure PGA | 0.1767 | 0.2518 | 0.3217 | 0.2691 | 0.1571 | 0.2353 |
| $D_{max}$ | HQGA | 0.0000 | 0.0000 | 0.0000 | 0.0000 | 0.0000 | 0.0000 |
| | pure PGA | 0.2701 | 0.3176 | 0.3918 | 0.3469 | 0.2284 | 0.3110 |
| $TS$ | HQGA | 1.0421 | 1.7975 | 1.9827 | 1.7469 | 1.4083 | 1.5955 |
| | pure PGA | 1.6112 | 1.3488 | 1.0199 | 0.9050 | 0.8220 | 1.1414 |
| $MS$ | HQGA | 0.8241 | 0.9087 | 0.8398 | 0.8230 | 0.7351 | 0.7182 |
| | pure PGA | 0.8249 | 0.7577 | 0.6570 | 0.7494 | 0.8771 | 0.7732 |
| $AQ$ | HQGA | 844.6961 | 814.8674 | 773.8195 | 858.0902 | 824.2611 | 823.1469 |
| | pure PGA | 881.3708 | 867.4635 | 816.9362 | 902.6367 | 859.5754 | 865.5965 |
| $t$ | HQGA | 27.4680 | 27.7650 | 27.4210 | 27.9530 | 27.7660 | 27.6746 |
| | pure PGA | 22.5930 | 22.5150 | 22.6090 | 22.7350 | 23.0310 | 22.6966 |

**Table 2.8.** Metric comparison for Scenario 4 when considering $f = (C_{max}, T_{max})$ (objective-based trimming)

| Metric | Method | Exp.1 | Exp.2 | Exp.3 | Exp.4 | Exp.5 | Average |
|---|---|---|---|---|---|---|---|
| $ONVG$ | HQGA | 25 | 21 | 17 | 13 | 19 | 19 |
| | pure PGA | 12 | 12 | 18 | 11 | 14 | 13 |
| $CM$ | HQGA | 0 | 0 | 0 | 0 | 0 | 0 |
| | pure PGA | 1 | 1 | 1 | 1 | 1 | 1 |
| $D_{av}$ | HQGA | 0.0000 | 0.0000 | 0.0000 | 0.0000 | 0.0000 | 0.0000 |
| | pure PGA | 0.2156 | 0.2438 | 0.1881 | 0.2765 | 0.2097 | 0.2267 |
| $D_{max}$ | HQGA | 0.0000 | 0.0000 | 0.0000 | 0.0000 | 0.0000 | 0.0000 |
| | pure PGA | 0.2855 | 0.3638 | 0.2213 | 0.3482 | 0.2303 | 0.2898 |
| $TS$ | HQGA | 1.4687 | 1.1218 | 1.1622 | 0.5541 | 0.8315 | 1.0276 |
| | pure PGA | 0.4884 | 0.4671 | 0.6024 | 0.9868 | 0.5446 | 0.6179 |
| $MS$ | HQGA | 0.7728 | 0.5464 | 0.4796 | 0.3464 | 0.3954 | 0.5081 |
| | pure PGA | 0.7323 | 0.8496 | 0.9116 | 0.8490 | 0.8643 | 0.8413 |
| $AQ$ | HQGA | 928.5704 | 853.6683 | 991.6876 | 937.6127 | 912.7221 | 924.8522 |
| | pure PGA | 980.2707 | 895.2686 | 1023.7922 | 980.1727 | 967.3411 | 969.3691 |
| $t$ | HQGA | 27.4060 | 28.3120 | 27.6400 | 27.2660 | 27.4370 | 27.6122 |
| | pure PGA | 22.4840 | 22.6100 | 22.6090 | 22.8590 | 22.5620 | 22.6248 |

exploration in discrete 0-1 hyperspace by using the updating operator of quantum gate as well as genetic operators of Q-bit, but the permutation-based GA was also applied for both exploration in permutation-based scheduling space and exploitation for good schedule solutions. The random key representation was used to convert the Q-bit representation to job permutation for evaluating the objective values of the schedule solution. In addition, to evaluate solution for multi-objective problems, a randomly weighted linear sum function was used in QGA, and a non-dominated sorting technique including classification of Pareto front and fitness assignment was applied in PGA regarding to both proximity and diversity. Particularly, to maintain diversity of population, two population trimming techniques were proposed and applied in the hybrid algorithm. Simulation results and comparisons with several performance metrics demonstrated the effectiveness and robustness of the proposed method for solving both single-objective and multi-objective PFSS problems. Apart from that, the algorithm can obtain solutions with good proximity and diversity for multi-objective problems.

As a novel intelligent computing technique, quantum-inspired evolutionary computation needs deep research and generalization, including theoretical, algorithmic and applied research work. As for QGA-based scheduling, we will study more effective and efficient hybrid frameworks to fuse QGA and other local searches, and develop some adaptive mechanisms to apply different trimming techniques, and propose more powerful way to handle multiple objectives as well. In addition, it is worth exploring the ability quantum-inspired evolutionary algorithm for other kinds of combinatorial optimization problems.

## Acknowledgement

This research is partially supported by National Science Foundation of China (60774082), National 863 Hi-Tech R&D Plan (2007AA04Z155) and the Project-sponsored by SRF for ROCS, SEM.

## References

1. Grover LK (1996) A fast quantum mechanical algorithm for database search. In: Proc. 28th ACM Symp. Theory of Computing, Pennsylvania:212–221
2. Shor PW (1994) Algorithms for quantum computation: discrete logarithms and factoring. In: Proc. 35th Symp. Foundation of Computer Science, Los Alamitos:20–22
3. Wang L (2001) Intelligence optimization with applications. Beijing: Tsinghua Univ. and Springer Press
4. Narayanan A, Moore M (1996) Quantum-inspired genetic algorithm. In: Proc. 1996 IEEE Int. Conf. Evolutionary Computation, Piscataway, NJ: IEEE Press:61–66
5. Han KH, Kim JH (2002) Quantum-inspired evolutionary algorithm for a class of combinatorial optimization. IEEE Trans. Evol. Comput., vol. 6, no. 6:580–593
6. Han KH, Kim JH (2004) A quantum-inspired evolutionary algorithm with a new termination criterion, gate, and two-phase scheme. IEEE Trans. Evol. Comput., vol. 8, no.2:156–169
7. Han KH, Kim JH (2000) Genetic quantum algorithm and its application to combinatorial optimization problem. In: Proc.2000 Congr. Evolutionary Computation, Piscataway, NJ: IEEE Press, vol. 2:1354–1360
8. Wang L (2003) Shop scheduling with genetic algorithms. Beijing: Tsinghua Univ. and Springer Press
9. Garey MR, Johonson DS (1979) Computers and intractability: a guide to the theory of NP-completeness. San Francisco: Freeman Press
10. Nawaz M, Enscore EJ, Ham I (1983) A heuristic algorithm for the m-machine, n-job flowshop sequencing problem. Omega, vol. 11:91–95

11. Ogbu FA, Smith DK (1990) Simulated annealing for the permutation flowshop problem. Omega, vol. 19:64–67
12. Wang L, Zhang L, Zheng DZ (2003) A class of order-based genetic algorithm for flow shop scheduling. Int. J. Adv. Manuf. Technol., vol. 22:828–835
13. Nowicki E, Smutnick C (1996) A fast tabu search algorithm for the permutation flow-shop problem. Eur. J. Oper. Res., vol. 91:160–176
14. Wang L, Zheng DZ (2003) An effective hybrid heuristic for flow shop scheduling. Int. J. Adv. Manuf. Technol., vol. 21:38–44
15. T'kindt V, Billaut JC (2002) Multicriteria Scheduling: Theory, Models and Algorithms. Berlin: Springer Press.
16. Silva JDL, Burke EK, Petrovic S (2004) An introduction to multiobjective metaheuristics for scheduling and timetabling. Lecture Notes in Economics and Mathematical Systems, vol. 535:91–129
17. Jaszkiewicz A (2002) Genetic local search for multi-objective combinatorial optimization. Eur. J. Oper. Res., vol. 137, no.1:50–71
18. Murata T, Ishibuchi H, Tanaka H (1996) Multiobjective genetic algorithm and its applications to flowshop scheduling. Comput. Ind. Eng., vol. 30, no. 4:957–968
19. Ishibuchi H, Murata T (1998) A multi-objective genetic local search algorithm and its application to flowshop scheduling. IEEE Trans. Syst. Man Cy. C.: Application and Reviews, vol. 28, no. 3:392–403
20. Murata T, Ishibuchi H, Gen M (2000) Cellular genetic local search for multi-objective optimization. In: Proc. 2000 Genetic and Evolutionary Computation Conf., Las Vegas, NV:307–314
21. Murata T, Ishibuchi H, Gen M (2001) Specification of genetic search directions in cellular multiobjective genetic algorithms. Lecture Notes in Computer Science:pp.82–95
22. Ishibuchi H, Yoshida T, Murata T (2002) Selection of initial solutions for local search in multiobjective genetic local search. In: Proc. 2002 on Congress Evolutionary Computation, Honolulu, HI:950–955
23. Ishibuchi H, Yoshida T, Murata T (2003) Balanced between genetic search and local search in memtic algorithms for multiobjective permutation flowshop scheduling. IEEE Trans. Evol. Comput., vol. 7, no. 2
24. Schaffer JD (1985) Multiple objective optimization with vector evaluated genetic algorithms. In: Proc. 1st Int. Conf. on Genetic Algorithms., Pittsburgh, PA:93–100
25. Arroyo JEC, Armentano VA (2005) Genetic local search for multi-objective flowshop scheduling problems. Eur. J. Oper. Res., vol. 167:717–738
26. Talbi H, Draa A, Batouche M (2004) A new quantum-inspired genetic algorithm for solving the traveling salesman problem. IEEE Int. Conf. Ind. Technol.:1192–1197
27. Wang L, Tang F, Wu H (2005) Hybrid genetic algorithm based on quantum computing for numerical optimization and parameter estimation. Appl. Math. Comput., vol. 171, no. 2:1141–1156
28. Bean JC (1994) Genetic algorithms and random keys for sequencing and optimization. ORSA Journal on Computing, vol. 6:154–160
29. Goldberg D (1989) Genetic algorithms in search, optimization, and machine learning. MA: Addison-Wesley.

30. Srinivas N, Deb K (1995) Multi-objective optimization function optimization using non-dominated sorting genetic algorithms. Evol. Comput., vol. 2, no. 3:221–448
31. Deb K, Pratap A, Agarwal A, Meyarivan T (2002) A fast and elitist mltiobjective genetic algorithm: NSGA-II. IEEE Trans. Evol. Comput., vol. 6, no. 2:182–197
32. Wolpert DH, Macready WG (1997) No free lunch theorems for optimization. IEEE Trans. Evol. Comput., vol. 1:67–82
33. Veldhuizen DAV (1999) Multiobjective evolutionary algorithms: classifications, analyses, and new innovations. Ph.D. dissertation, Department of Electrical and Computer Engineering, Graduate School of Engineering, Air Force Institute of Technology, Wright-Patterson AFB, Ohio
34. Zitzler E (1999) Evolutionary algorithms for multiobjective optimization: methods and applications. Ph.D. dissertation, Swiss Federal Institute of Technology (ETH), Zurich, Switzerland
35. Czyak P, Jaskiewicz A (1998) Pareto simulated annealing – a metaheuristic technique for multiple objective combinatorial optimization. Journal of Multi-Criteria Decision Making, vol. 7:34–47
36. Ulungu EL, Teghem J, Ost C (1998) Efficiency of interactive multi-objective simulated annealing through a case study. J. Oper. Res. Soc., vol. 49:1044–1050
37. Tan KC, Goh CK, Yang YJ, Lee TH (2006) Evolving better population distribution and exploration in evolutionary multi-objective optimization. Eur. J. Oper. Res., vol. 171:463–495
38. Jaszkiewicz A (2003) Do multiple-objective metaheuristcs deliver on their promises? A computational experiment on the set-covering problem. IEEE Trans. Evol. Comput., vol. 7, no. 2:133–143
39. Carlier J (1978) Ordonnancements a contraintes disjonctives. R.A.I.R.O. Recherche operationelle/Oper. Res., vol. 12:333–351
40. Reeves CR (1995) A genetic algorithm for flowshop sequencing. Comput. Oper. Res., vol. 22:5–13
41. Nelson RT, Sarin RK, Daniels RL (1986) Scheduling with multiple performances measures: the one machine case Manage. Sci., vol. 2:464–479
42. Shanthikumar JG (1983) Scheduling n jobs on one machine to minimize the maximum tardiness with minimum number tardy jobs. Comput. Oper. Res., vol. 10, no. 3:255–266
43. Selen WJ, Hott D (1986) A mixed integer goal programming formulation of the standard flow shop scheduling problem. J. Oper. Res. Soc., vol. 37, no. 12:1121–1126
44. Ho JC, Chang YL (1991) A new heuristic for n-job, m-shop problem. Eur. J. Oper. Res., vol. 52:194–202
45. Rajendran C (1992) Two-stage flowshop scheduling problem with bicriteria. J. Oper. Res. Soc., vol. 43, no. 9:871–884
46. Liao CJ, Yu WC, Joe CB (1997) Bicriteria scheduling in the two machines flow shop. J. Oper. Res. Soc., vol. 48:929–935
47. Loukil T, Teghem J, Tuyttens D (2005) Solving multi-objective production scheduling problems using metaheuristics. Eur. J. Oper. Res., vol. 161:42–61
48. Armento VA, Ronconi DP (1999) Tabu search for total tardiness minimization in flowshop scheduling problems. Comput. Oper. Res., vol. 26:219–235
49. Potts CN, Wassenhove LNV (1982) A decomposition algorithm for the single machine total tardiness problem. Oper. Res. Lett., vol. 1:177–181

50. Taillard E (1993) Benchmarks for basic scheduling problems. Eur. J. Oper. Res., vol. 64:278–285

# 3

# Quantum Simulataneous Recurrent Networks for Content Addressable Memory

Raheel Allauddin[1], Stuart Boehmer[2], Elizabeth C. Behrman[2], Kavitha Gaddam[3], and James E. Steck[4]

[1] Department of Computer Science, Wichita State University, Wichita, KS 67260-0083 USA
[2] Department of Physics, Wichita State University, Wichita, KS 67260-0032 USA
elizabeth.behrman@wichita.edu
[3] Department of Mechanical Engineering, Wichita State University, Wichita, KS 67260-0133 USA
[4] Department of Aerospace Engineering, Wichita State University, Wichita, KS 67260-0044 USA

**Summary.** We explore a model for a quantum Hopfield artificial neural network, in which qubits are prepared in an initial state and allowed to evolve to steady state. We derive the equations for finding their stable states, by minimizing the Lyapunov energy. We apply the equations first to the cytoskeleton of a biological neuron, where the qubits are microtubulins in a microtubule. Our method can reproduce classical calculations made earlier by Tuszynski *et al.*, but because it is a fully quantum method can also explore the possibility that an artificial computing structure based on an array of tubulins can act as a quantum computer. We then derive a method for training the quantum network to converge to a stable target pattern, given an input pattern, and show that a spatial array of qubits can be trained to perform the CNOT, which with a rotation is a universal quantum gate. This means that a large enough array can in principle be trained to do any calculation.

## 3.1 Introduction

In today's world of digital computers, all information processing and storage are encoded in the form of discrete values, known as bits, represented by some sort of physical system. In the early nineteenth century, mechanical computers used the position of gear teeth as a bit in a complicated gear train system. Currently the most popular electronic physical system possesses two clearly identifiable states called digital bits representing logic values 0 and 1, and all information storage, manipulation and retrieval are in this form. Certain quantum systems are able to encode bits as physical states. For example when we measure the spin of an electron, it is always in one of two possible states; we can define one state as "up" or $|\uparrow\rangle$ and the other as "down" or $|\downarrow\rangle$. In

R. Allauddin et al.: *Quantum Simulataneous Recurrent Networks for Content Addressable Memory*, Studies in Computational Intelligence (SCI) **121**, 57–76 (2008)
www.springerlink.com 

other words $|\uparrow\rangle$ means that the spin was found to be parallel to the axis along which the spin was measured; $|\downarrow\rangle$ means that the spin was found to be anti-parallel to the axis of the measurement. If a bit is represented by a two level quantum mechanical system then we call such a bit a quantum bit or qubit. A qubit is governed by the laws of quantum physics; thus it can be in both "up" and "down" states simultaneously, unlike the digital bit that can be in only one state at a time. These in between states are called superposition states and are basically a blend of some or all-possible states simultaneously. Thus for a qubit, a superposition state can be of the form $C_0 |\uparrow\rangle + C_1 |\downarrow\rangle$, where $C_0$ and $C_1$ are two complex numbers such that $|C_0|^2 + |C_1|^2 = 1$. When physically measured, the quantum mechanical superposition state collapses to one of the two possible states: $|\uparrow\rangle$ with probability $|C_0|^2$ or $|\downarrow\rangle$ and with probability $|C_1|^2$. When a quantum system is used to implement a computer, the quantum superposition of qubits allows the possibility of parallel data storage and quantum parallel calculations. When a quantum system is used to implement a neural network, these same advantages extend the capability of network calculation and storage.

The introduction of neural networks into the field of quantum computing (QC) results in quantum neural networks (QNNs) which combine the advantages of both QC and ANNs. Several different approaches have been suggested [1–6].

A quantum Hopfield network is a fully inter-connected network where each neuron is connected to other neurons in the network. Hopfield networks [7–9] are capable of storing information and pattern completion when used as an "associative memory" or "content addressable" memory, serve as a pattern recognition system, and solve optimization problems. The network is initialized to a particular state and each neuron is updated according to its weighted inputs until it converges to a stable state, which is a local minimum in the Lyapunov energy of the network. The stable state is a pattern that is stored and can be recalled by the network.

In section 3.2 we introduce the Hopfield model of associative memory and explain the function of the neural network. In sections 3.3 and 3.4 we define our proposed quantum Hopfield network and show how it works in a similar fashion to the classical Hopfield neural network, though with greatly increased storage capacity due to its ability to store superposition states. Section 3.5 consists of an application of QHN to microtubules, which are a structural component of biological neurons in the brain, and have been proposed [10,11] as a possible way in which the brain could be performing quantum processing. It is found that quantum information processing in microtubules is feasible, though at temperatures (5.8 K) lower than physiological temperatures. In section 3.6 it is shown that quantum Hopfield networks can be trained using simultaneous recurrent backpropagation [12] to give us a Quantum Simultaneous Recurrent Network, or QSRN. In this section an algorithm is developed to train the QHN to recognize a pattern. We demonstrate the algorithm by showing that the QSRN can be trained to do a CNOT in section 3.7.

## 3.2 The Hopfield Model

The Hopfield neural network is one model of associative content–addressable memory; it is capable of storing information as well as carrying out certain computational tasks such as pattern completion or error correction. In Hopfield neural networks the processing units are fully interconnected, *i.e.*, each unit is connected to every other unit. Hopfield [7] demonstrated that associative memory can be implemented with his network and optimization tasks such as the traveling salesman problem can be solved. A Hopfield network is initialized to a beginning state. Then each neuron's output is updated according to its weighted inputs. The updating of the neurons continues until the network converges to a stable state, or fixed point.

A *binary* Hopfield network has neurons which can be in one of two states: $S_i = +1$ (on) and $S_i = -1$(off), where $i$ is the number of the neuron. These are exactly analogous to bits 0 and 1. Each neuron is synaptically connected to other neurons with multiplicative weights which are real numbers. Each neuron is randomly selected, sums up its input, and decides whether to turn itself on or off. Let $W_{ij}$ be the synaptic strength from neuron '$j$' to neuron '$i$' and let $I_i$ be the external input to unit '$i$'. Each neuron takes a weighted sum of its inputs according to the following equation:

$$S_i = \mathrm{sgn}\left(\sum_{k\neq i} W_{ij}S_j + I_i\right) . \tag{3.1}$$

After repeated updating, the network may achieve a stable point, meaning, in the state space, all neurons remain in their current state after examining the inputs. The output is a stable vector-of-states which constitutes the recall of a stored pattern in the memory.

In the *continuous* Hopfield net, updates in time are continuous, rather than discrete. The governing equation of the $i^{th}$ neuron is

$$\frac{du_i}{dt} = \sum_j W_{ij}S_j - \frac{u_i}{R_i} + I_i \ , \tag{3.2}$$

where

$$S_j = \tanh(\gamma u_j) = f(u_j). \tag{3.3}$$

$R_i$ is the decay rate ($R_i > 0$) and $S_j$ is the output of unit $j$ after the activation function $f$ is applied. $\gamma$ is the gain of the activation function. The Hopfield network considered in this paper is in the high gain limit where $\gamma$ and $R_i$ are large, and the tanh approximates the sgn function of the discrete network.

Each state in the high gain Hopfield neural network has a Lyapunov energy function, given by

$$L = -\frac{1}{2}\sum_i \sum_j W_{ij} S_j S_i - \sum_j S_j I_j \; . \tag{3.4}$$

$L$ is a Lyapunov function for the dynamics of equations 3.2 since it is bounded and it can be shown [7] that $\frac{dL}{dt} < 0$ except at the fixed points. Thus the Hopfield network is globally asymptotically stable and the stable states are the minima of the "energy" function $L$. The processing of the Hopfield neural network, from an initial to a final steady state, is a process by which the "energy" of the system becomes smaller and eventually achieves a local minimum at the stable state. The weights, the external inputs, and the initial state determine which stable state, or pattern, is recalled by the network.

## 3.3 THE QUANTUM SYSTEM

A two-level quantum system can be put into the Hopfield form to obtain the quantum Hopfield net (QHN) as follows. Consider an array of N qubits. We write the Hamiltonian, or energy function, for each qubit $j$ as

$$H_J = K_j \sigma_{xj} + A_j \sigma_{zj} \; ; \tag{3.5}$$

where $\sigma_x$ and $\sigma_z$ are the Pauli matrices, and the full Hamiltonian, for the whole system, is $H = \sum_j H_j$. The first term represents the flipping of the qubit from one state to the other, with amplitude $K_j$; the second, the energy difference $2A_j$ between the two states. This difference can be the result of external fields or interactions with other qubits.

For the whole system we need to sum over all the qubits. For a system at finite temperature we write the partition function $Q = tr \; \exp(-\beta H)$, where $tr$ is the trace and $\beta$ the inverse temperature in units of Boltzmann's constant. The partition function contains information about all possible states of the quantum system, and how they depend functionally on each variable and each parameter. This will lead us to our definition of the quantum Hopfield net. Following [13], and using Kac's replica method [15], we write the trace in the basis set of state variables (eigenfunctions of $\sigma_z$) $\{S_j = \pm 1\}$, and use the Feynman path integral [16] formulation:

$$Q = \int \prod_{j=1}^N \mathcal{D}\left[S_j(t)\right] \exp\{ -\int_0^\beta dt \sum_{j=1}^N K\sigma_{xj}(t) + \int_0^\beta dt \int_0^\beta dt' \sum_{j \neq j'=1}^N \sigma_{zj}(t)\alpha_{jj'}\sigma_{zj'}(t') \}; \tag{3.6}$$

where the matrix $\alpha(t,t')$ comes from the second term in equation 3.5, and contains both interactions between qubits and between qubits and external fields. The symbol $\mathcal{D}$ indicates the summation over all possible paths (the "path integral".) A quantum mechanical particle (or system of particles) does not follow, as a classical system does, one minimum action path (trajectory) between specified initial and final points, but instead all possible paths; this is one way of understanding "quantum dispersion", or the "fuzziness" of a quantum state, the fact that a quantum particle does not in general have a definite state in the particular basis considered. The fuzziness is represented by the spatial extent of the loops, as we will see below. Here $\alpha$ represents the interaction between qubits, and the state of the qubit is represented by the variables $\{S_j = \pm 1\}$, the eigenvalues of the operator $\sigma_z$ for each qubit $j$. The path integral formulation is sometimes referred to [13] as propagation in imaginary time, because the Boltzmann factor $\exp(-\beta H)$ is the same as the expression for time evolution $\exp(-itH/\hbar)$, for $t = -i\beta\hbar$, where $\hbar$ is Planck's constant divided by $2\pi$. (Note that the Wick rotation of the time evolution operator $\exp(-itH/\hbar)$ to the Boltzmann factor $\exp(-\beta H)$ changes a unitary operator into a non-unitary one; this creates some formal difficulties and necessitates care. These issues have been discussed in the literature. See [17] for a pedagogical account, and references cited therein.)

The first term in the exponential in equation 3.6 is the tunneling term, containing the amplitude for tunneling, $K$, multiplied by the Pauli matrix $\sigma_x$. The second term contains the interactions between each pair of qubits $j, j'$, with an interaction strength $\alpha_{jj'}$ which may, in general, depend on both the identities of the interacting qubits and their separation in imaginary time. Because the partition function is a trace, the state of each qubit at "time" 0 is constrained to be equal to its state at "time" $\beta$, so the path for each qubit can be thought of as a loop. (This periodicity in imaginary time is one of the consequences mentioned in the previous paragraph.) The path integral (enumeration of all possible paths the system can follow) is equivalent to an infinite number of sums (over the intermediate states) as follows:

$$Q = \lim_{n\to\infty} \textstyle\sum_{S_j=\pm 1} \exp\{\ \xi \displaystyle\sum_{j=1}^{N}\sum_{t=0}^{n-1} S_j(t)S_j(t+1) \\ + \sum_{j\neq j'=1}^{N}\ \sum_{t,t'=0}^{n-1} S_j(t)\alpha_{jj'}(t,t')S_{j'}(t')\ \}; \quad (3.7)$$

where the functions $\{S_j\}$ are the instantaneous values assumed by each of the qubits $j$, at each imaginary time $t$ and $t'$ (such that $0 < t, t' < \beta$ ) along the inverse temperature loop, and the first term accounts for the tunneling of the qubit, where

$$\xi = -\frac{1}{2}\ln\left[\tanh(\beta K/n)\right]\ . \quad (3.8)$$

For full details see [1, 13, 16]. The discretized representation is exact in the limit that $n$ is infinite. The continuum of possible states of each qubit is represented by a continuum of bits. For finite n we can compute the state of the net exactly by summing over all possible states of each of the discretized points of each of the qubits.

Qualitatively, the imaginary time (inverse temperature) loops represent the quantum dispersion, or spreading out of the wave function. At low temperature (the quantum limit) $\beta$ is large, so each qubit has a great deal of freedom (a long imaginary time) before it is constrained to return to its original point. In this limit a very large number n will be necessary to represent the dispersion. At higher temperature $\beta$ becomes smaller, until in the classical limit the dispersion loop shrinks to a single point (*i.e.* $n = 1$.) Physically the qubit is no longer able to tunnel. The (finite) value of n we use will be large enough (the approximation will be good) if $\beta H/n$ is small.

## 3.4 Mapping the Quantum System onto the Hopfield Net

Each set of instantaneous values of each of the qubits, $\{S_j(t)\}$, can be thought of as a microstate. The contributions from all of the possible microstates are summed over to obtain the partition function. The contribution from a microstate to the partition function is large if its associated energy is low. Thus, states that are more probable can be found by maximizing the argument of the exponential in equation 3.7 or, since each exponential corresponds to a Boltzmann factor $\exp(-\beta E)$, minimizing its associated energy. We rearrange that argument to correspond with equation 3.4; that is, we define the Lyapunov function for the QHN as the integral of the Hamiltonian over the imaginary time loop *i.e.*, the quantity in curly brackets in equation 3.7 - divided by $\beta$:

$$L_{QHN} = -\frac{1}{\beta}\Big[ \quad \xi \sum_{j=1}^{N} \sum_{t=0}^{n-1} S_j(t) S_j(t+1) + \sum_{j \neq j'=1}^{N} \sum_{t,t'=0}^{n-1} S_j(t) \alpha_{jj'}(t,t') S_{j'}(t') \Big] . \tag{3.9}$$

This is the QHN. Classically, the high gain Hopfield net replaces a continuum of possible values of each neuron with a finite number; in the QHN it becomes again a continuum, as described below.

Information in the Hopfield net is stored as stable states. Classically the discrete net converges to the corners of an $N$-dimensional box [7], where the neuron outputs necessarily equal plus or minus 1. In a network of $N$ classical neurons there are $2^N$ possible states (corners of the box.) Once we discretize the quantum Hopfield net to discretization number $n$, there are $2^{N\times n}$ possible states: that is, each corner of the $N$-dimensional classical Hopfield box has $2^n$ states. The greater the discretization $n$, the greater the number of possible states. When the temperature is lower we need a larger discretization number;

thus, the lower the temperature the greater the storage capacity of the QHN. For infinite $n$ we have the fully quantum mechanical system, and have a fully faithful representation of the quantum dispersion - all possible superposition states.

Consider a QHN at finite $n$, bearing in mind that only $n = 1$ (classical) and $n = \infty$ (fully quantum mechanical) have any recoverable physical meaning; still, "$n = \infty$" should be interpreted as equivalent to $\beta H/n \ll 1$, which can always be satisfied with finite n. (That is, given a physical system with Hamiltonian $H$, at an inverse temperature $\beta$, there is an $n$ such that $\beta H/n$ is much less than 1.) Figure 1 shows a picture of the net for two qubits ($N = 2$) and two discretization points ($n = 2$.) Each loop represents a qubit or two-level system (TLS), propagating in imaginary time (inverse temperature), from $t = 0$ to $t = \beta$.

Each dot on a loop represents the instantaneous state of the qubit at a particular value of $t$, and these dots interact with their nearest neighbors along the loop (the first term in equation 3.9.) This is the tunneling term. The arrows show the direction of integration, from $t = 0$ to $t = \beta$. Each dot also interacts with other dots on other loops according to the second term in equation 3.9. In the simplest possible model, which we consider here, the two qubits 1 and 2 interact only at equal imaginary times (*i.e.*, $t = t'$), and the Lyapunov function from equation 3.9 becomes:

$$L_{QHN} = -\frac{1}{\beta}\left[\xi\left(\sum_{t=0}^{n-1} S_t S_{t+1} + \sum_{t=0}^{n-1} S'_t S'_{t+1}\right) + \alpha \sum_{t=0}^{n-1} S_t S'_t\right], \tag{3.10}$$

where the (primed, unprimed) S functions refer to the two interacting loops.

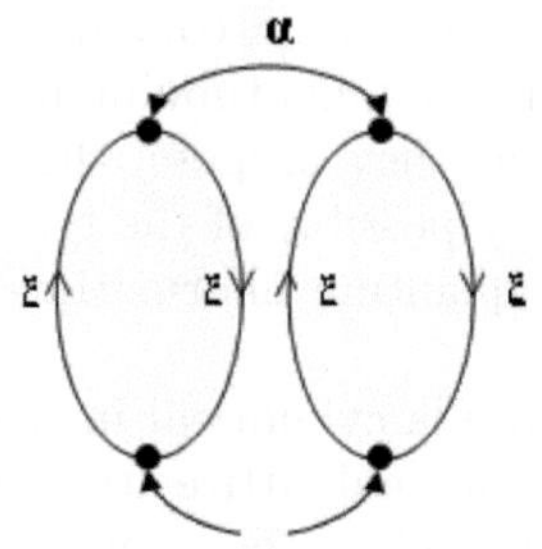


**Fig. 3.1.** Representation of two interacting qubits, with discretization $n = 2$. Each loop represents a qubit, propagating in imaginary time (inverse temperature) from 0 to $\beta$. The interactions between discretization points on the same loop have a strength of $\xi$; the (bidirectional) interactions between points on different loops, of $\alpha$.

As mentioned earlier all the extrema are in the corners. These can be minima, maxima or saddle points. The Hopfield algorithm guarantees a minimum, or at least a relative minimum (subject to the constraint that each $S$ has absolute value 1.) We can see that this is true by putting equations 3.2 and 3.4 together, to obtain

$$\frac{dL}{dt} = -\sum_{i}^{n} \frac{dS_i}{du_i} \left( \frac{du_i}{dt} \right)^2 \leq 0 \,, \tag{3.11}$$

since the tanh is a strictly increasing function. We can analyze the Lagrangian in equation 3.9 to identify the stable points of the quantum Hopfield net. We do this by minimizing $L$ subject to the constraint that each $S$ must have absolute magnitude less than or equal to one using the method of Lagrange multipliers [18]. (The tanh substitution does this automatically for the Hopfield net.) We can define a new functional

$$F \equiv L + \sum_{i}^{n} l_i h_i \,, \tag{3.12}$$

where $h_i = (S_i^2 - 1)$ and $l_i$ are the Lagrange multipliers, each multiplying a quantity $h_i$ that is constrained to be less than or equal to zero. The Kuhn-Tucker condition [18] then tells us that there is a vector $l$ of multipliers, such that $\nabla F = 0$ and $l \cdot h = 0$, where $h$ is the vector of conditions and $\nabla$ is as usual the gradient (with respect to each of the $S$ variables).

## 3.5 QHN As An Information Propagator for a Microtubulin Architecture

Recently several researchers [10, 11] have proposed that the microtubules of neurons in the brain act as a quantum computer. The idea is that a microtubule is composed of an array of microtubulin dimers, each of which can have two polarization states. If, as they propose, these dimers act as qubits, the microtubule is a spatial array possibly of the type we have been considering, so that we can analyze its quantum information processing using the QHN model [14].

The microtubule structure is cylindrical in shape consisting of a number of tubulins arranged in a hexagonal lattice structure, as shown in Figure 3.2. The tubulin dimers are oriented as they have a net electronegative charge towards the alpha-monomer. There are 13 tubulins along the circumference of the microtubule. Each tubulin has six neighbors and interacts with its six neighbors. There can be propagation of information from one qubit to another in the network due to the (Coulombic) interactions between each pair of qubits. The output is calculated as the lowest energy state of the qubit. Figure 3.3 shows the dipole-dipole interactions of a single tubulin with its neighbors. The "+" sign represents a parallel arrangement and the "-" sign represents an anti-parallel arrangements of dipole moments.

We can apply our QHN methodology to this model of the array of microtubulins. The values of intra- and inter-molecular distances, length of microtubule (based on the number of rows), tunneling amplitude, temperature, external field, leakage term, gain parameter, etc., are set, using experimental values where possible. The QHN is started out at some initial state after assigning values to all the parameters in the algorithm, by assigning an intial dipole value (orientation of the polarization) to each qubit. The output of each neuron is updated according to its weighted inputs. The updating procedure of the network minimizes the Hopfield energy or Lyapunov function and continues until the network converges to a stable state achieving a local minimum. This stable state is a pattern that is recalled by the network. This final state is the output of the network.

**Fig. 3.2.** Structure of a microtubule and the arrangement of its tubulins.

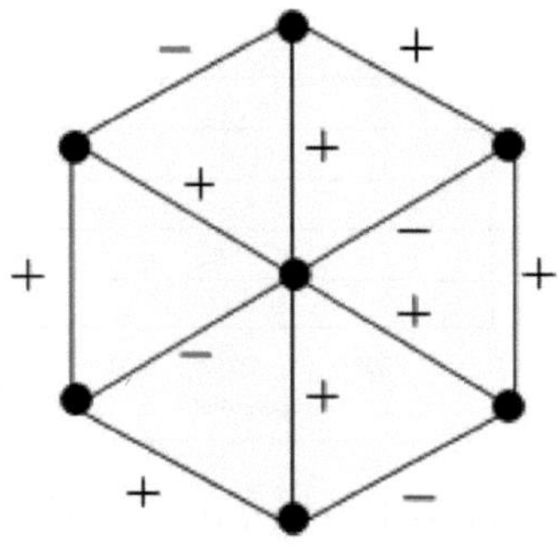

**Fig. 3.3.** Dipole-dipole interactions showing parallel or anti-parallel arrangements of dipole moments.

To test the validity of our method we first compare it to some published classical calculations on this model. Tuszynski, *et al.* [10,11], have performed some Monte Carlo computations to observe the size dependence of the phase behavior in microtubules. Figures 3.4 and 3.6 are reproduced from that paper. Figures 3.4 and 3.5 show the mean polarization per site for a triangular lattice whose size is 13 x 26 (26 layers, 13 qubits in each row, 338 qubits), in the classical limit ($n = 1$). The initial values of spins Figure 3.5 were taken from Tuszynski [10]. It can be seen that the QHN results are comparable to the Monte Carlo calculations. For the large lattice limit, Tuszynski has a calculation of 5000 layers; due to computational capacity problems we were only able to do 3000 layers; nonetheless our results are comparable to theirs, as seen in Figures 3.6 and 3.7. Again, Figures 3.6 and 3.7 are in the classical limit.

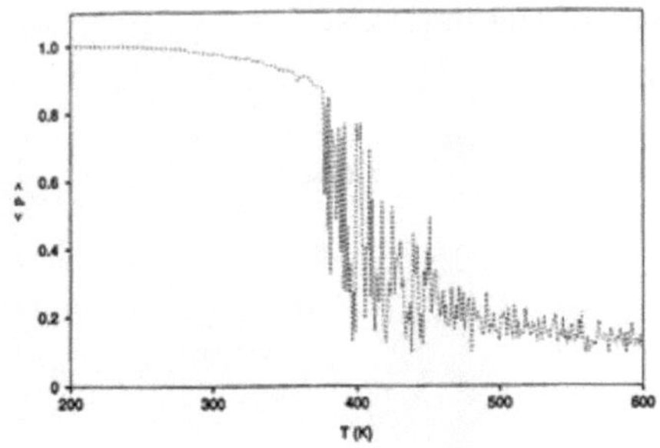


**Fig. 3.4.** Mean Polarization per layer, as a function of temperature, for a 13 x 26 lattice, from ref. [10] as calculated by classical Monte Carlo.

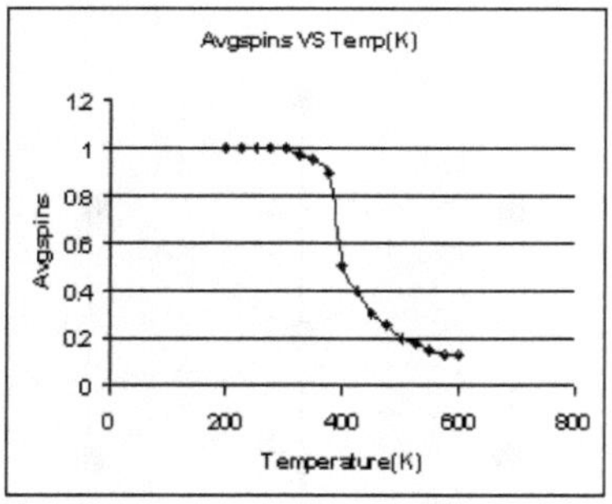


**Fig. 3.5.** Average polarization as function of temperature, for a 13 x 26 lattice, using values for the parameters of ref. [10], calculated by QHN, in the classical limit ($n = 1$).

With some confidence in our method we now move to nonzero quantum mechanical effects, setting $n > 1$. In order that our discretization be valid we require $\beta E/n < 1$, where $E$ is the energy of the system. Because of computational limitations we used only 26 layer system, or 338 qubits. All other

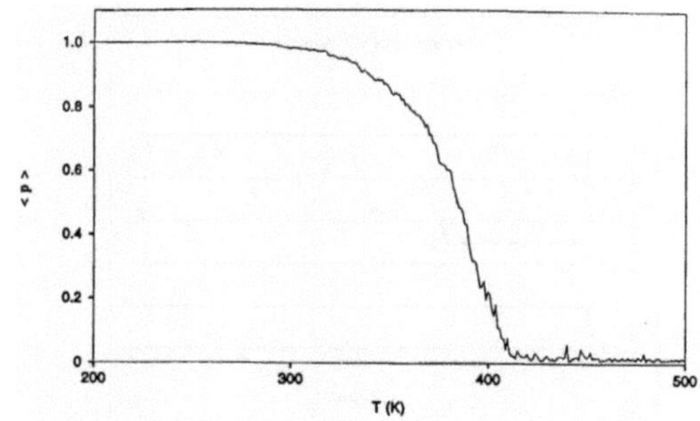


**Fig. 3.6.** The mean polarization for a 13 x 5000 lattice, as calculated by classical Monte Carlo (ref. [10])

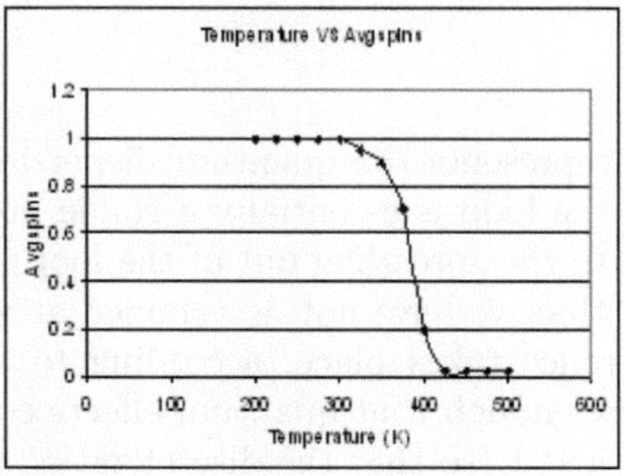


**Fig. 3.7.** Average polarization as a function of temperature for a 13 x 3000 lattice, calculated by QHN, in the classical limit ($n = 1$).

experimental parameters were kept as before, and we look for local minima for the QHN.

Some interesting phenomena were observed. An example is shown in Figure 3.8, where we see the variation of average spin along the layers of the microtubule, at a temperature of 5.8 K ($\beta = 2$ MeV), starting from an initialization of the qubits in a superposition state of 1 down and 3 up spins. The QHN was run for 100,000 epochs, at which the configuration was stable.

It can be observed from the graph that there is a locally stable configuration in which a nearly constant polarization at one end is transformed to a different one at the other, a kind of information processing effect. That is, we have found a stable configuration for which the "input" at one end of the microtubule is transformed to an "output" at the other, and both input and output are superposition states (*i.e.*, the average spin (polarization) is between minus one and one. Since superposition states cannot be realized classically, the microtubule is acting as a quantum computer, with fully quantum mechanical inputs and outputs. This is of course lower than physiological temperatures, and indeed no similar "processing" effect was observed at temperatures as high as those in a living brain: the higher the temperature, the smaller the

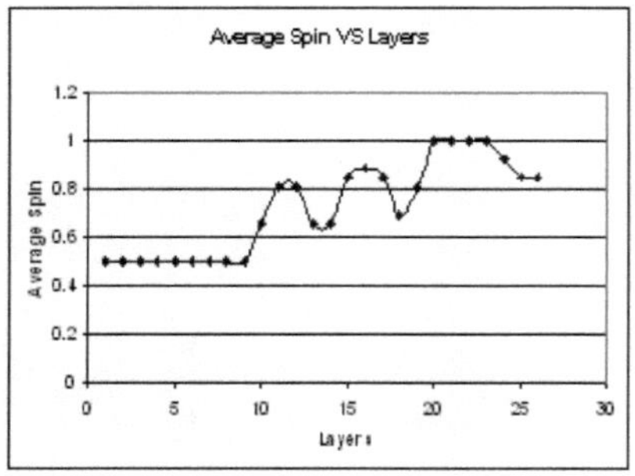


**Fig. 3.8.** The variation of average spins along the layers of the 13 x 26 microtubule for $n = 4$ with an initial configuration of 1 down, 3 up spins for each qubit, as calculated by QHN, at $T$=5.8 K. This final configuration was stable at 100,000 passes.

quantum loop, which represents the quantum dispersion, shrinks. At physiological temperatures the loop is essentially a single point, and no quantum effects (which are due to the spreading out of the loop) can be observed.

It should be noted that we have not determined at what temperature the loss of quantum coherence takes place, according to our model: 5.6K was used because it was low enough that quantum effects could be observed, and high enough that $\beta E/n < 1$ (so that the discretization $n = 4$ was sufficient.) We would expect the coherence temperature to be a function of the size of the system and of its interactions; it is possible that Nature cleverly uses "noiseless" or "error avoiding" quantum codes and/or subsystems, such as have been proposed to make artificial quantum computing practical [19]. Further work in this area would be of value. In particular, one could look at the dispersion (size of the loop) as a function of temperature.

## 3.6 QHN As Simultaneous Recurrent Network

In this section we show that we can train a QHN network to the target pattern by manipulating the external parameters. Hopfield networks act as associative memories [7–9], and can work as simultaneous recurrent networks, or SRNs [12], which can solve more complex problems.

Let us consider a rectangular array of qubits, of size p×q. The dynamics of QSRN from equation 3.2 is given by

$$\frac{\partial u_j^i(t)}{\partial t} = H_j^i(\{w\}, S(t)) \; ; \tag{3.13}$$

where $i$ is the qubit index, $j$ is the loop discretization point and $H_j^i(\{w\}, S(t))$ is given by

$$H_j^i(\{w\}, S(t)) = -\frac{1}{2}\ln\left[\tanh\left(\frac{\beta K_i}{n}\right)\right]\left[S_{j+1}^i(t) + S_{j-1}^i(t)\right]$$
$$+\varepsilon_i + \sum_{k\neq i=1}^{p\times q} \chi_{ik} S_j^k(t) \; ; \tag{3.14}$$

where we have put the Lyapunov function from eqs. 3.4 and 3.9 into eq. 3.2, and where we have split the general interaction term $\alpha$ (from eqs. 3.6, 3.7, and 3.9) into a self term (external field) of strength $\varepsilon$, and an interaction term $\chi$, *i.e.*, $\alpha_{jj'} = \varepsilon_j \delta_{jj'} + \chi_{jj'}$ . Note that $H_j^i(\{w\}, S(t))$ is a function of the set of weights $\{w\}$ and of all the spins $\{S\}$, which are functions of their Hopfield evolution "time". (It is important to note that Hopfield evolution is not real time evolution. The Hopfield equations find local minima, but the updating process that leads to these minima does not mimic the actual behavior of the system in real time. Hopfield "time" is also distinct from what we are calling "imaginary time" – that is, inverse temperature. The inverse temperature of a spin variable does not change in our calculations: we station different virtual spins along the imaginary time loop, and they stay there.) We want to minimize the output error subject to the constraint that the given dynamics hold. Following the method of [12], we have

$$L = \sum_i \frac{1}{2}\left[D^i - S_{avg}^i(t_f)\right]^2$$
$$+\sum_{i,j}\int_0^{t_f} \lambda_j^i(t)\left[\frac{\partial u_j^i(t)}{\partial t} - H_j^i(\{w\}, S(t))\right] dt \; ; \tag{3.15}$$

where $D^i$ is the $i^{th}$ desired value for the average of $S^i$ around its loop, $\lambda_j^i(t)$ are the Lagrange multipliers for the $j^{th}$ discretization point of the $i^{th}$ qubit at Hopfield time $t$, and $t_f$ is the final Hopfield time. Integrating by parts on the first term in the integral gives us

$$L = \sum_i \frac{1}{2}\left[D^i - S_{avg}^i(t_f)\right]^2 + \sum_{i,j}\left[\lambda_j^i(t) u_j^i(t)\right]_0^{t_f}$$
$$+\sum_{i,j}\int_0^{t_f}\left[u_j^i(t)\frac{\partial \lambda_j^i(t)}{\partial t} - \lambda_j^i(t) H_j^i(\{w\}, S(t))\right] dt \,. \tag{3.16}$$

Taking the variation of $L$ with respect to $u$, $\lambda$, and $w$, we have:

$$
\begin{aligned}
\delta L = &-\sum_i \left[D^i - S^i_{avg}(t_f)\right] \frac{1}{n} \sum_{j=1}^{n} f'\left(u^i_j(t_f)\right) \delta u^i_j(t_f) \\
&+\sum_{i,j} \left[\delta\lambda^i_j(t) u^i_j(t) + \lambda^i_j(t)\delta u^i_j(t)\right]_0^{t_f} \\
&+\sum_{i,j} \int_0^{t_f} \left\{ \left[ -\delta u^i_j(t) \frac{\partial \lambda^i_j(t)}{\partial t} - u^i_j(t) \frac{\partial \delta\lambda^i_j(t)}{\partial t} - \delta\lambda^i_j(t) H^i_j(\{w\}, S(t)) \right] \right. \\
&\left. +\lambda^i_j(t) \left( \frac{\partial H^i_j(t)}{\partial w} \delta w - \sum_{k,l} \frac{\partial H^i_j(t)}{\partial S^k_l} f'(u^k_l(t) \delta u^k_l(t)) \right) \right\} dt \; ; \qquad (3.17)
\end{aligned}
$$

where $f$ is the activation function from equation 3.3, and $f'$ its derivative with respect to $u$. At $0 < t < t_f$ setting the terms combining $\delta u^m_n(t)$ equal to zero gives

$$
0 = -\frac{\partial \lambda^m_n(t)}{\partial t} - f'\left(u^m_n(t)\right) \sum_{i,j} \lambda^i_j(t) \frac{\partial H^i_j(t)}{\partial S^m_n} \; . \qquad (3.18)
$$

Equation 3.18 defines back propagation through time (backward: $t_f \to 0$) for the Lagrange multipliers $\lambda^i_j$. At $t = t_f$ setting the terms multiplying $\delta u^m_n(t)$ equal to zero gives

$$
\left[D^i - S^i_{avg}(t_f)\right] \frac{1}{n} f'\left(u^m_n(t_f)\right) + \lambda^m_n(t_f) = 0 \; . \qquad (3.19)
$$

This is the error defined at $t = t_f$ ; $u^m_n(0)$ is specified as an initial condition, so $\delta u^m_n(0) = 0$. Setting the terms combining $\delta w$ equal to zero does not give a solution for the weights, so the weight updates are done by gradient descent:

$$
w^{new} = w^{old} + \eta \frac{\partial L}{\partial w} \; , \qquad (3.20)
$$

where $\eta$ is the learning rate, and where

$$
\frac{\partial L}{\partial w} = -\sum_{i,j} \int_0^{t_f} \frac{\partial H^i_j(t)}{\partial w} \lambda^i_j(t) \, dt \; . \qquad (3.21)
$$

Since we only know $S$ and $u$ at the final time $t_f$ , but not at any intermediate times, we must approximate. We choose as follows: At $t = t_f$,

$$
\lambda^m_n(t_f) = \left[D^n - S^m_{avg}(t_f)\right] \frac{1}{n} \left[f'\left(u^m_n(t_f)\right)\right] \; . \qquad (3.22)
$$

We now solve the differential equation backward in time, but using only the spins and states at $t_f$:

$$
\frac{\partial \lambda^m_n}{\partial t} = -f'\left(u^m_n(t_f)\right) \sum_{i,j} \lambda^i_j(t) \frac{\partial H^i_j(t_f)}{\partial S^m_n} \; ; \qquad (3.23)
$$

with

$$w^{new} = w^{old} + \eta \frac{\partial H_j^i(t_f)}{\partial w} \int_0^{t_f} \lambda_j^i(t)dt \; ; \tag{3.24}$$

for each weight $w$. These are the training equations for the quantum simultaneous recurrent network.

It is important to note, as stated earlier, that these equations we have derived do not represent the actual time evolution of the quantum system. Rather, they enable our designing parameters for a quantum system that will have energy minima at configurations that we can specify. That is, the training tells us the values the parameters need to have in order that the system have energy minima in the specific states we desire. Physical implementation of the QSRN would involve something like an annealing process, to induce the system to relax to the desired energy minimum. (Thus, the Hamiltonian of the system in such a physical implementation would not be the one above, and, indeed, would be time dependent.)

## 3.7 Application: The CNOT Gate

We now apply our proposed QSRN algorithm, developed in section 3.6, to a spatial matrix of qubits. These could be quantum dots, or SQUIDs, or something else. The idea is to make something like a microchip of qubits that can be trained to different desired functions. The quantum mechanical training parameters under consideration are tunneling amplitudes $K$, field strengths $\varepsilon$, and coupling strengths $\chi$. These parameters are the adjustable weights of the network, and are repeatedly updated in the direction of decrease in error of the system.

As the algorithm in section 3.6 suggests, we set the initial spins of points in the input layer to "up" (computationally we map "up" as $0.99 \approx 1$ and "down" as $-0.99 \approx -1$) The initial states of the output qubits and of any qubits in hidden layers, are set to zero; that is, half the discretization points are initialized as up and the other half as down, so the average spin of the qubit in the output layer is zero. This value of zero represents complete quantum uncertainty as to the state of the qubit. We then let the network evolve until it settles into some local minimum of the energy functional. At this point the output (average state over the temperature loop) is calculated and compared with the target value (desired value), the error is calculated, and the training weights are adjusted.

It is well known [20] that a controlled-NOT, or CNOT, in combination with a rotation, is universal for quantum computing. We therefore choose this gate as a useful representative example of the training of the QHN. This gate involves a set of four pairs of inputs with associated outputs: the state of the target qubit is flipped iff the state of the control qubit is up.

For this gate we found we needed an intermediate layer ("hidden layer") consisting of two qubits. Thus we had six qubits in two horizontal rows, which we label by row: 1 through 3 for the top row, and 4 through 6 for the bottom. Thus the leftmost column is the input layer (numbers 1 and 4), the values of whose discretization points are initialized as either all up or all down; the middle column is the hidden layer (numbers 2 and 5); and the rightmost column is the output layer (numbers 3 and 6) whose averages are read out. See Figure 3.9. Both the hidden layer points and the output layer points were initialized to zero (by setting half up and half down: complete quantum uncertainty as to the state.) With a time step of 0.0025 (*i.e.*, $1.7 \times 10^{-15}s$, in units of Planck's constant $h$) it took 40 or 50 steps to attain equilibrium (*i.e.*, a state in which all $S_{ij}$ for a given qubit were near either +1 or -1). A graph of the typical behavior of $S_6$ over Hopfield "time" is given in Figure 3.10.

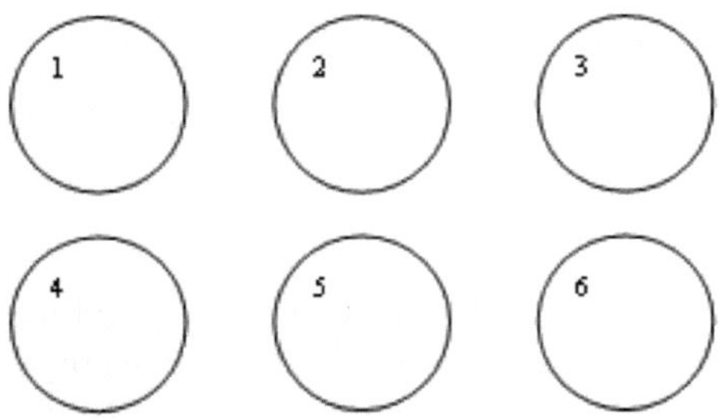


**Fig. 3.9.** Array of qubits used for the calculation of the CNOT gate. Each circle represents the imaginary time (inverse temperature) loop of a single qubit, whose average properties we calculate approximately with a finite discretization. Each qubit interacts with all other qubits at equal imaginary times (qubit-qubit coupling $\chi$) and with an external field $\varepsilon$. The tunneling term $K$ acts within each loop. Qubits 1 and 4 comprise the input layer; qubits 3 and 6, the output layer; and qubits 2 and 5, the hidden layer.

In order to find parameters $(K, \varepsilon, \chi)$ that would produce the desired equilibrium (*i.e.*, with outputs corresponding to the given inputs for a CNOT) we found it was necessary to use small learning rates ($\eta_K = 0.00001$, $\eta_\xi = 0.001$, and $\eta_\varepsilon = 0.0001$.) Error as a function of epoch is shown in Figure 3.11. Note that by 5 epochs the error is essentially zero. The final solution (values for the training weights) is given in Tables 3.1 and 3.2.

Discretization error was checked directly by repeating the calculation several times, doubling the number of points each time: $n = 2, 4, 8, 16$, and so on. We found that the solution (calculated weights) was stable by $n = 4$. We also checked that the solution was stable as a function of decreasing temperature: Increasing $\beta$ (decreasing $T$), while keeping $\beta/n$ constant, showed this down to a temperature of $T = 19$ K.

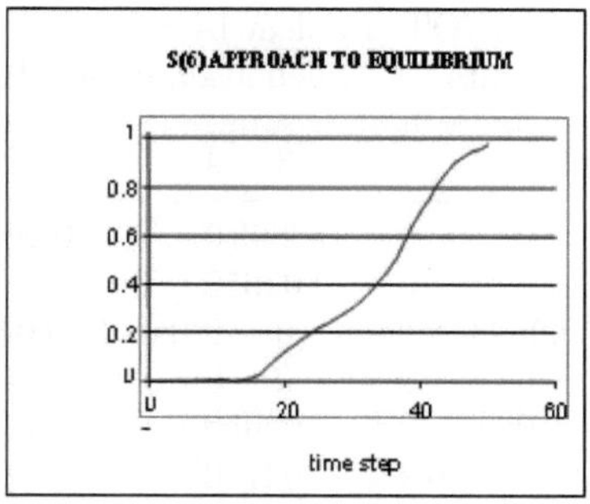


**Fig. 3.10.** Approach of the average value of an output qubit, $S_6$, to equilibrium with advance through Hopfield "time."

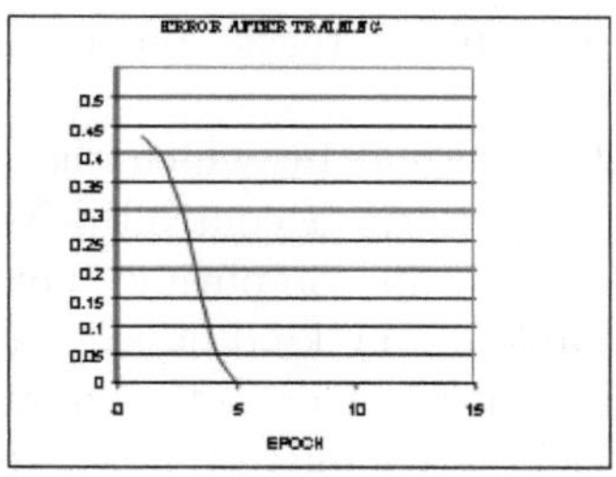


**Fig. 3.11.** Error as a function of epoch, in the training of a six-qubit array to the CNOT gate. Calculation parameters are as given in Table 3.1, and calculated final weights in Tables 3.1 and 3.2.

**Table 3.1.** Solution of the CNOT problem by a $2 \times 3$ array of qubits: tunneling amplitudes $K$ and external fields $\varepsilon$, for each qubit number $i$. Learning rates were: $\eta_K = 10^{-7}$, $\eta_{coul} = 10^{-5}$, $\eta_\varepsilon = 10^{-6}$, inverse temperature $\beta = 0.7$, and gain $\gamma = 10$. Discretization number used was $n = 8$.

| i | 1 | 2 | 3 | 4 | 5 | 6 |
|---|---|---|---|---|---|---|
| $K_i$ | 0.1 | 0.1 | 0.1 | 0.1 | 0.1 | 0.0733 |
| $\varepsilon_i$ | 0.1 | 0.1 | 0.0801 | 0.1 | 0.1 | 0.0142 |

## 3.8 Summary

We have presented a model for a quantum Hopfield neural network, and shown how it evolves from an initialized state to a final equilibrium state. We have demonstrated that the net is quantum mechanical by exploring a physical implementation (the microtubule), for which we have shown that the net can take one superposition state to another. This shows a kind of information processing that is intrinsically quantum mechanical, that is, that is impossible for a classical computer, neural or algorithmic, since a classical computer cannot be

**Table 3.2.** Solution of the CNOT problem by a 2x3 array of qubits: qubit-qubit coupling $\chi$ for each pair of qubits $i, j$. Calculation parameters are given in Table 3.1.

| i/j | 1 | 2 | 3 | 4 | 5 | 6 |
|---|---|---|---|---|---|---|
| 1 | 0 | 0 | 0.0893 | 0 | 0 | 0.0019 |
| 2 | 0 | 0 | -0.0265 | 0 | 0 | -0.133 |
| 3 | 0.0893 | -0.0265 | 0 | 0 | -0.0265 | -0.0474 |
| 4 | 0 | 0 | 0 | 0 | 0 | 0 |
| 5 | 0 | 0 | -0.0265 | 0 | 0 | -0.133 |
| 6 | 0.0019 | -0.133 | -0.0474 | 0 | -0.133 | 0 |

put into a superposition state. It is ability to manipulate superposition states that makes possible Grover's algorithm [21] for rapid data base searches. Even more interesting would be manipulation of entanglement, which makes possible Shor's algorithm [22] for rapid factorization. Work demonstrating learning of these is in progress.

We have also derived a method by which the net can be trained, and demonstrated its ability to perform a controlled-NOT, which (with a pure rotation) is universal for quantum computing. Thus, in principle, we have shown that this net is capable of any kind of calculation.

The QHN is a simultaneous recurrent net, a quantum SRN. According to Paul Werbos [12] SRNs offer a completely new avenue for researchers. We believe ours is the first application of this idea to the quantum realm, and we think that this idea offers many possibilities that should be explored. Design of algorithms that take advantage of quantum superposition and quantum entanglement turns out to be very difficult: Despite a great deal of work over the past decade by a large number of very smart people, few have been generated. It is not at all clear that a way will be or can be found to generate algorithms efficiently to solve general problems on quantum computers. This is all the more reason to explore a quantum *neural* approach: Perhaps a net can learn what we cannot design or have difficulty in designing.

In this first, proof-of-concept paper, we have demonstrated both quantum processing (section 3.5) and learning of a benchmark universal gate (section 3.7); we have also derived a general learning technique. In addition, the quantum Hopfield model provides a way actually to analyze numerically the possibility of quantum computing by biological systems, about which there has been a great deal of speculation for many years. We are currently working on ways by which to ameliorate the exponential difficulty of simulation so as to be able to apply our methods to larger problems like signal and image processing, for which there may be a quantum advantage.

## 3.9 Acknowledgements

This work was supported by the National Science Foundation, Grants ECS-9820606 and 0201995. We thank J.A. Tuszynski for permission to reproduce Figures 4 and 6 from his published work. We also thank Anil Javali and Preethika Kumar Gagnebin for helpful discussions.

## References

1. Behrman E C, Nash L R, Steck J E, Chandrashekar V G, Skinner S R (2000) Information Sciences 128:257-269
2. DeGaris H, Sriram R, Zhang Z (2003) Proceedings of the International Joint Conference on Neural Networks 4:2589-2593
3. Bandyopadhyay S, Menon L, Kouklin N, Williams P F, Ianno N J (2002) Smart Mater. Struc. 11:761-766
4. Ezhov A (2003) Proceedings of SPIE 5128:107-119
5. Resconi G, van del Wal A J (2002)Information Sciences 142:249-273
6. Allauddin R, Gaddam K, Behrman E C, Steck J E, Skinner S R (2002) Proceedings of the International Joint Conference on Neural Networks 3:2732-2737
7. Hopfield J J (1982) Proceedings of the National Academy of Sciences USA 79: 2554-2558
8. Haykin S (1999) Neural Networks: a Comprehensive Foundation. Prentice Hall, New York
9. Abu-Mostafa Y S, St. Jacques J-M (1985) IEEE Trans. Information Theory IT31:461-464
10. Tuszynski J A, Hameroff S, Hurylak P (1998) Phil Trans R. Soc. Lond. A 356:1897-1926
11. Hagan S, Hameroff S R, Tuszynski J A (2002) Phys. Rev. E 65:061901-11
12. Werbos P (1997) Handbook of Neural Computation, release 97/1, A2.3:2
13. Behrman E C, Jongeward G A, Wolynes P G (1983) J. Chem. Phys. 79:6277-6281
14. Behrman E C, Gaddam K, Steck J E, Skinner S R Microtubulins as a quantum Hopfield network. In Tuszynski JA (ed.) The Emerging Physics of Consciousness. Springer Verlag, 2006. This paper was written after, but published before, the present article, and contains an expanded and extended version of the method and results in section 3.5. Though published subsequently, we consider the present work to be the foundational paper.
15. Kac M (1968) Mathematical mechanisms of phase transitions. In Chretien M, Gross E, Deser S (eds.) Statistical Physics: Phase transitions and superfluidity. 1:241-305
16. Feynman R P, Hibbs A R (1965) Quantum Mechanics and Path Integrals. McGraw-Hill, New York
17. Polyak M (2004) arXiv: math.GT/0406251
18. Luenberger D G (1984) Linear and Nonlinear Programming. Addison-Wesley, Reading, MA
19. Zanardi P, Rasetti M (1997) arXiv: quant-ph/9705044
20. Nielsen M A, Chuang I L (2000) Quantum computation and quantum information. Cambridge University Press, Cambridge, UK

21. Grover L K (1996) Proceedings of the 28th Annual ACM Symposium on Theory of Computing. ACM, New York, pp.212-9.
22. Shor P W (1997) SIAM J. Computing 26: 1484

# 4

# Quantum Intelligent Mobile System

Chunlin Chen[1] and Daoyi Dong[2]

[1] Department of Control and System Engineering, Nanjing University, Nanjing 210093, China clchen@nju.edu.cn
[2] Key Laboratory of Systems and Control, Institute of Systems Science, AMSS, Chinese Academy of Sciences, Beijing 100080, China dydong@amss.ac.cn

A brand-new paradigm of intelligent systems–quantum intelligent mobile system–is proposed through the fusion of quantum technology with mobile system. A quantum intelligent mobile system (QIMS) is essentially a complex quantum-classical hybrid autonomous system which generally consists of four fundamental components: quantum information processing units (QIPU), multi-sensor system, controller/actuator, and quantum/classical information convertors. A hybrid architecture based on multi-quantum-agent system is proposed for specific requirements of this intelligent mobile system, and a multi-sensor system is designed with SQUID sensor and quantum well Hall sensor, where quantum sensors coexist with traditional sensors. According to the requirements of certain tasks and hardware performance, Grover algorithm is presented for searching problem in path planning, and the theoretic result shows that it can reduce the problem complexity of $O(N^2)$ in traditional intelligent mobile system to $O(N^{3/2})$. Then a novel quantum reinforcement learning (QRL) algorithm is proposed for the quantum intelligent mobile system and a learning example demonstrates the validity and superiority of QRL. This quantum intelligent mobile system has many potential applications in various areas and also offers a platform for the research on quantum or quantum-inspired technologies.

## 4.1 Introduction

Intelligent mobile systems (robots) are a kind of mobile systems which have the ability to accomplish certain tasks through perceiving environments with sensors and acting upon those environments with effectors. Related research fields, such as mobile robotics, have been very important research areas in artificial intelligence and control theory for decades. Researchers have developed series of methodologies to build mobile systems with self-learning and adaptive control capabilities, ranging from purely software mobile systems to hybrid systems with hardware. With the development of research in this

C. Chen and D. Dong: *Quantum Intelligent Mobile System*, Studies in Computational Intelligence (SCI) **121**, 77–102 (2008)
www.springerlink.com 

promising area, more challenging capabilities are realized through advanced technologies.

As viewed from the trend of mobile systems, increasing intelligence and reducing the physical size of systems are two important research and development thrusts. The key to increasing the intelligence of a system lies in improving the performance of its physical apparatus and speeding up learning, planning and decision making. At the same time, the reduction of physical size will cause quantum effects to become more prominent. Hence many new challenges must be addressed from new angles. In this chapter, we consider the role of quantum mechanics in intelligent mobile systems.

In the last century, quantum mechanics is one of the most important achievements. In fact, it provides a very useful mathematical framework for the construction of physical theories and the development of new technologies. In 1970s, some techniques for controlling single quantum systems began to develop rapidly and the applications of quantum theories gradually interested many scientists. In these applications, quantum information technology is the most attractive field due to the potential capability of parallel computation and secret communication [1], [2]. In 1980s, some original ideas about quantum information technology were proposed. For example, Wootters and Zurek discovered that an arbitrary unknown quantum state can not be precisely cloned [3]. Benioff first proposed a quantum mechanical computer model based on Turing machines [4] and simultaneously Feynman observed that quantum computers consisting of quantum-mechanical systems have an information-processing capability much greater than that of corresponding classical systems [5], [6]. In 1985 Deutsch described a quantum Turing machine and showed that quantum computers could indeed be constructed [7]. Bennett and Brassard proposed a protocol using quantum mechanics to distribute keys between two parties without any possibility of a compromise [8].

Since the 1990s quantum information technology has rapidly developed. In the aspect of quantum computation, two important quantum algorithms, Shor algorithm [9], [10] and Grover algorithm [11], [12] have been proposed in 1994 and 1996, respectively. Shor algorithm can give an exponential speedup for factoring large integers into prime numbers and it has been realized for the factorization of integer 15 using nuclear magnetic resonance (NMR) [13]. Grover algorithm can achieve a square speedup over classical algorithms in unsorted database searching and its experimental implementations have also been demonstrated using NMR [14]-[16] and quantum optics [17], [18] for a system with four states. Moreover some other quantum computing tasks, such as quantum games [19]-[21] and Deutsch-Jozsa algorithm [22], are also realized using NMR. For quantum communication, some important advancements have been made. For example, Bennett and co-workers proposed a teleportation scheme where an unknown quantum state can be disembodiedly transported to a desired receiver through purely classical communication and purely nonclassical Einstein-Podolsky-Rosen (EPR) correlations [23]. Experimental demonstration of quantum teleportation has been realized with quantum optics [24], [25] and NMR [26]. Some entanglement swapping [27] and entanglement purification [28] experiments have also been realized. Experimental entanglement of five- and six-photons has been demonstrated [29], [30].

In recent years, the powerful potential of quantum information has also drawn the attention of scientists in artificial intelligence (AI). For example, Benioff first proposed the concept of quantum robot in 1998 , where a quantum robot is described as a mobile quantum system that includes an on-board quantum computer and needed ancillary systems [31]. Recently Dong and co-workers have proposed a brand-new quantum robot from an engineering perspective and consider its information exchanges and learning control [32]. The quantum computing version of artificial neural network has also been studied from pure theory to simple simulated and experimental implementation [33]-[36]. Rigatos and Tzafestas use quantum computation for the parallelization of a fuzzy logic control algorithm to speed up the fuzzy inference [37]. And quantum or quantum-inspired evolutionary algorithms are proposed to improve the existing evolutionary algorithms [38], [39]. Taking advantage of quantum computation, the algorithm integration inspired by quantum characteristics will not only improve the performance of existing algorithms on traditional computers, but also promote the development of related research areas such as quantum computer and machine learning. Considering the essence of computation and algorithms, Dong and his co-workers have presented the concept of quantum reinforcement learning (QRL) inspired by the state superposition principle and quantum parallel computation [40] and have demonstrated its advantages and feasibility through simulated experiments [41], [42]. Following this concept, we in this chapter propose a formal framework of quantum intelligent mobile system (QIMS) and demonstrate some related methodologies, algorithms and applications.

The organization of this chapter is as follows. Section 4.2 gives a brief introduction to intelligent mobile system and quantum mechanics. In section 4.3, after the presentation of intelligent agent in quantum system, a systematic architecture for quantum mobile system is proposed based on multi-quantum-agent. The hardware of typical QIMS is discussed in section 4.4. Section 4.5 focuses on planning and related planning methods for QIMS are presented based on the Grover algorithm. The theoretical results show that QIMS can reduce the complexity of the planning problem from $O(N^2)$ to $O(N^{3/2})$. Section 4.6 focuses on learning for QIMS and a novel quantum reinforcement learning (QRL) algorithm is proposed based on quantum superposition and quantum parallelism. Simulation results demonstrate that QIMS is superior to classical mobile system in efficient learning under the QRL algorithm. Finally, section 4.7 and 4.8 present some discussions and conclusions.

## 4.2 Prerequisite

Now we know that quantum intelligent mobile system (QIMS) is an interdiscipline of mobile system and quantum mechanics. Hence in this section we firstly give a brief introduction to these two areas.

### 4.2.1 Intelligent mobile system

Mobile system is a general area including autonomous vehicles, mobile robots, softbots etc. In this chapter we will be concerned primarily with intelligent mobile robots and the intelligent mobile system is studied from the following three aspects: architecture, physical implementation, planning and learning.

Architecture is the foundation of designing the whole system and different architectures have been used as the software and hardware framework of mobile robots, such as the Sense-Model-Plan-Act approach and behavior-based systems [44]. One recent trend in mobile robotic architectures has been a focus on hybrid architecture which combined both of the above approaches [45]. No mater what kind of architecture is selected, the intelligent mobile systems should acquire the following capabilities: (1) Adaptive learning. While the system expanding its capabilities, the architecture can evolve by learning without modifying code by people. (2) Scalable ability. Every part of the architecture is autonomous and integrated. So it is easy to add new parts and functions for new purpose, which makes the whole architecture scalable and compatible with an increment of the mobile system capabilities. (3) Efficient and universal communication.

Physical implementation focuses on the hardware and software of the whole quantum intelligent mobile system. As for the quantum intelligent mobile system, we know that most parts of the system are quantum-classical hybrid ones.

Planning and learning are the key functions of intelligent system. In QIMS, these two functions may be implemented using quantum computation and quantum algorithms, which play very important roles in speeding up learning and planning.

### 4.2.2 Fundamentals of quantum mechanics

Quantum mechanics is a mathematical framework and it is built on several postulates. Hence we briefly introduce these postulates before we come to the formal framework of QIMS. Since we mainly concern the information processing capability of QIMS, we often explain these postulates with quantum bit (qubit).

**Postulate 1** *The state of a closed quantum system is represented by a complex vector space with inner product (i.e. Hilbert space) known as the state space of the system. The system is completely described by its state vector, which is a unit vector in the system's state space.*

In quantum mechanics, the quantum state $|\psi\rangle$ (Dirac representation) in a Hilbert space is also called the wave function for this quantum system. The inner product of two wave functions $|\psi_1\rangle$ and $|\psi_2\rangle$ can be written into $\langle\psi_1|\psi_2\rangle$ and the normalization condition for wave function $|\psi\rangle$ is $\langle\psi|\psi\rangle = 1$. As the simplest quantum mechanical system, a qubit has a two-dimensional state space and its two basic states are denoted as $|0\rangle$ and $|1\rangle$. Different from classical bit, a qubit can also lie in both $|0\rangle$ and $|1\rangle$ at the same time besides $|0\rangle$ or $|1\rangle$. In the other words, a qubit $|\psi\rangle$ can generally be expressed as a linear combination of $|0\rangle$ and $|1\rangle$

$$|\psi\rangle = \alpha|0\rangle + \beta|1\rangle \tag{4.1}$$

where $\alpha$ and $\beta$ are complex coefficients. This special quantum phenomenon is called state superposition principle, which is an important difference between classical computation and quantum computation [2]. $\alpha$ and $\beta$ are also called probability amplitudes. The magnitude and argument of probability amplitude represent amplitude and phase. Its normalization condition is equivalent to $|\alpha|^2 + |\beta|^2 = 1$.

**Postulate 2** *The time evolution of the state of a closed quantum system is described by the Schrödinger equation*

$$i\hbar\frac{d|\psi\rangle}{dt} = H|\psi\rangle \tag{4.2}$$

*In this equation, $\hbar$ is Planck's constant and $H$ is the Hamiltonian of the closed system.*

The evolution process in Postulate 2 can be described by a unitary transformation. If the state of system at time $t_1$ is $|\psi_1\rangle$ and the state at time $t_2$ is $|\psi_2\rangle$, the evolution process from $t_1$ to $t_2$ can be described by a unitary operator $U(t_1, t_2)$ which depends only on the times $t_1$ and $t_2$

$$|\psi_2\rangle = U(t_1, t_2)|\psi_1\rangle \tag{4.3}$$

In fact the procedure of quantum information processing can be represented by a series of unitary transformations, which can be accomplished by some basic quantum logic gates such as quantum NOT gate, Hadamard gate, phase gate, $\pi/8$ gate and CNOT gate. The quantum NOT gate can be represented as follows

$$X \equiv \begin{pmatrix} 0 & 1 \\ 1 & 0 \end{pmatrix} \tag{4.4}$$

It takes the quantum state $\alpha|0\rangle + \beta|1\rangle$ to $\beta|0\rangle + \alpha|1\rangle$. Hadamard gate (or Hadamard transform) is one of the most useful quantum gates and can be represented as [1]:

$$H \equiv \frac{1}{\sqrt{2}}\begin{pmatrix} 1 & 1 \\ 1 & -1 \end{pmatrix} \tag{4.5}$$

The phase gate $S$ and $\pi/8$ gate $T$ can be represented, respectively

$$S \equiv \begin{pmatrix} 1 & 0 \\ 0 & i \end{pmatrix} \tag{4.6}$$

$$T \equiv \begin{pmatrix} 1 & 0 \\ 0 & exp(i\pi/8) \end{pmatrix} \tag{4.7}$$

The above quantum gates are single qubit gates. The most important two-qubit gate is controlled not (CNOT) gate $U_{CNOT}$

$$U_{CNOT} = \begin{pmatrix} 1 & 0 & 0 & 0 \\ 0 & 1 & 0 & 0 \\ 0 & 0 & 0 & 1 \\ 0 & 0 & 1 & 0 \end{pmatrix} \tag{4.8}$$

The CNOT gate can complete the task 'If $A$ is true, then do $B$'. That is, if the control qubit is set to $|1\rangle$ then the target qubit is flipped, otherwise the target qubit is left alone. In fact, the Hadamard gate, phase gate, $\pi/8$ gate and CNOT gate are universal for quantum computation and the unitary transformations in quantum information processing can be approximated using their combination.

Now let's see what will happen when we observe a quantum system. Postulate 3 can describe the effects of measurements

**Postulate 3** *Quantum measurements are described by a collection $\{M_m\}$ of measurement operators. These are operators acting on the state space of the system being measured. The index $m$ refers to the measurement outcomes that may occur in the experiment. If the state of the quantum system is $|\psi\rangle$ immediately before the measurement then the probability that result $m$ occurs is given by*

$$p(m) = \langle\psi|M_m^\dagger M_m|\psi\rangle \tag{4.9}$$

*and the state of the system after the measurement is*

$$|\psi'\rangle = \frac{M_m|\psi\rangle}{\sqrt{\langle\psi|M_m^\dagger M_m|\psi\rangle}} \tag{4.10}$$

*The measurement operators satisfy the completeness equation*

$$\sum_m M_m^\dagger M_m = I \tag{4.11}$$

When we measure a quantum system in a superposition state, it will collapse to one of its eigenstates. If the system measured is a qubit in superposition state $|\psi\rangle$, the qubit system would collapse into either $|0\rangle$ or $|1\rangle$. However, we can not determine in advance whether it will collapse to state $|0\rangle$ or $|1\rangle$. We only know that we get this qubit in state $|0\rangle$ with probability $|\alpha|^2$, or in state $|1\rangle$ with probability $|\beta|^2$.

**Postulate 4** *The state space of a composite quantum system is the tensor product of the component quantum systems. If we have systems numbered* 1 *through* $n$*, and system number* $i$ *is prepared in the state* $|\psi_i\rangle$*, then the joint state* $|\psi\rangle$ *of the total system can be represented as:*

$$|\phi\rangle = |\psi_1\rangle \otimes |\psi_2\rangle \otimes \ldots |\psi_n\rangle \tag{4.12}$$

*where '*$\otimes$*' means tensor product.*

Consider an $n$-qubit system, which can be looked upon as a composite quantum system of $n$ idependent qubits and its quantum state can be represented as

$$|\phi\rangle = \sum_{x=00\ldots0}^{11\ldots1} C_x|x\rangle \tag{4.13}$$

where $\sum_{x=00\ldots0}^{11\ldots1} |C_x|^2 = 1$. $|x\rangle$ can take on $2^n$ values, so the superposition state can be looked upon as the superposition of all integers from 0 to $2^n - 1$. It is valuable to note that if a bipartite quantum state $|\psi\rangle_{AB}$ can not be written into the tensor product of $|\psi\rangle_A$ and $|\psi\rangle_B$, we call the quantum system lies in an entangled state. Quantum entanglement is a peculiar phenomenon which has no classical corresponding and it is becoming an important physical resource and has been used for quantum communication, quantum error-correction and quantum cryptography [1], [2].

## 4.3 Architecture of QIMS based on multi-quantum-agent

### 4.3.1 Intelligent agent in quantum system

Intelligent agent is a new concept framework in AI and control theory. An agent is anything that can be viewed as perceiving its environment through sensors and acting upon that environment through effectors. There are mainly four types of agents: (1) Simple reflex agents; (2) Agents that keep track of the world; (3) Goal-based agents; (4) Utility-based agents.

With the development of quantum computation and quantum communication, the research of quantum control theory and quantum intelligent information processing has interested most researchers. Here the concept of agent is considered in quantum systems as an important component and structure for quantum control and information processing systems, which makes the foundation for quantum algorithms and quantum intelligent systems and also builds a bridge between quantum computation and traditional artificial intelligence.

An agent in quantum system can be looked upon as an quantum agent. Different from traditional agent, the state of the quantum agent can lie in a superposition state and quantum agents can be taken as the generalization of the traditional agent concept.

Here we use a modified structure called utility-based learning agents to give a unified view for the basic components of the mobile system. The overall structure is shown in Fig.4.1.

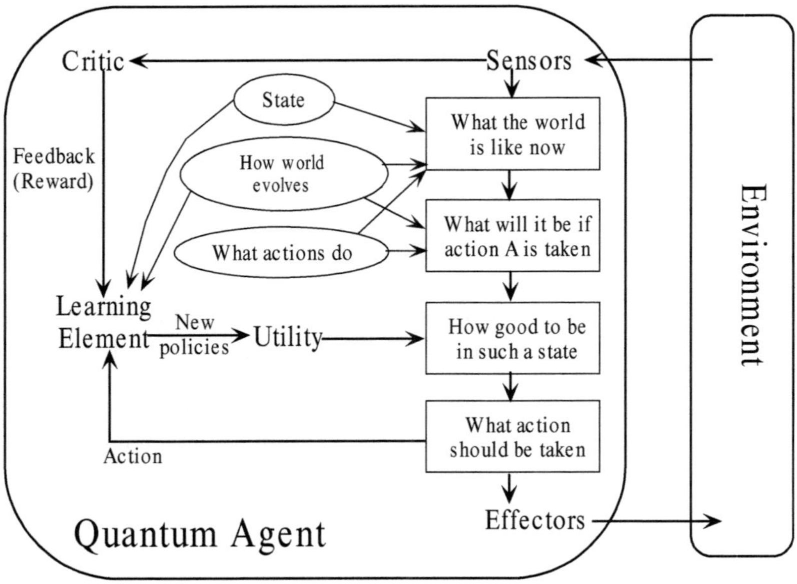


**Fig. 4.1.** Structure of utility-based learning agent

In this structure utility is a function that maps a state onto a real number, which describes the associated degree of happiness. The learning element makes the agent smarter through learning experiences. Fig.4.1 shows an integrated structure of the utility-based learning agent, however, not all of the components are prerequisite when implementing these agents. From this structure we can not see any difference between traditional agent and quantum agent, which means the concept of agent is compatible with quantum agent. The difference lies in that each entities of a quantum agent may be quantum or quantum-classical hybrid ones.

### 4.3.2 Multi-quantum-agent system

Analog to the multiagent system, some domains also require multi-quantum-agent system (MQAS). The systems may be composed of homogeneous non-communicating quantum agents, heterogeneous non-communicating quantum agents or heterogeneous communicating quantum agents. The categories of heterogeneous communicating quantum agents are the most general ones. The most distinct characteristic of the multi-quantum-agent systems is the way how the quantum agents communicate with each other, which is very different from the communication between traditional agents.

Different quantum agents can interact with each other and exchange information through quantum bus. The quantum bus may be a refreshable entanglement resource [43] or some other quantum circuits. Besides general functions of a classical bus such as CAN bus, PCI bus and PC104 bus, the quantum bus should also be able to exchange or preprocess quantum information. The advantages of quantum communication such as high channel capability, perfect security can be fully utilized. According to quantum information theory, the quantum agents can exchange information with each other more rapidly and secretly than traditional agents. For example, different quantum agents can accomplish entanglement swapping and quantum teleportation.

Entanglement swapping can entangle two quantum agents that never directly interact [27]. Consider two pairs of entangled quantum agents AB and CD, and initially A, B are entangled and C, D are also entangled. Now we make appropriate projective measurement on agent B and agent C, such as Bell-state measurement, and let agent B and agent C entangle each other. Without any operation on agent A and agent D or direct interaction between A and D, the joint measurement on BC automatically makes agents A and D collapse into an entangled state, that is to say, two pairs of entangled agents AB and CD perform entanglement swapping and they become two entangled pairs BC and AD. Entanglement swapping makes it possible that different quantum agents can communicate with each other without direct interaction.

Quantum teleportation is a process, in which an unknown quantum state can be disembodiedly transported to a desired receiver through purely classical communication and purely nonclassical Einstein-Podolsky-Rosen (EPR) correlations. A sender (agent Alice) is given an unknown state and also has one of two particles prepared in an EPR state. She performs a Bell-state measurement on the composite system of the unknown state and her EPR particle, and informs the receiver (agnet Bob) the result of this measurement through classical communication. Knowing this, Bob performs one of four possible unitary transformations on the second EPR particle and converts it into an exact replica of the unknown state. Through quantum teleportation, a quantum agent can send disembodiedly its information to remote another quantum agent.

### 4.3.3 Architecture of QIMS based on MQAS

As the design framework of QIMS, architecture connects every module of the whole system and is critical for the implementation of hardware and software. There are mainly three kinds of architectures for a traditional mobile system: deliberate, reactive [44] and hybrid architecture [45]. In recent years, hybrid architecture is the focus of mobile system and it can solve the modelling problem in unknown environment with high real time. At the same time, hybrid architecture is able to represent the environment reasonably, which can't be achieved by single architecture. As for QIMS, we also choose hybrid architecture and according to the characteristics of QIMS, a hybrid architecture based on MQAS is proposed. Besides the elementary traits of being autonomous, intelligent and real-time, it is also excellent in the following ways: (1) harmonizing deliberate modules and reactive modules and making it possible to acquire reactive ability as well as long time planning ability; (2) bearing strong self-learning ability so that the system can improve by itself through learning; (3) scalable and new modules can be fused with original system.

Hence we design three kinds of agents: (1) I/O agents, including sensor agents and actuator agents; (2) deliberate agents; (3) reactive agents. To share relative information and knowledge, a blackboard system which consists of memory parts and reasoning parts is used to organize different information and make decisions. The framework of the whole architecture is shown in Fig.4.2.

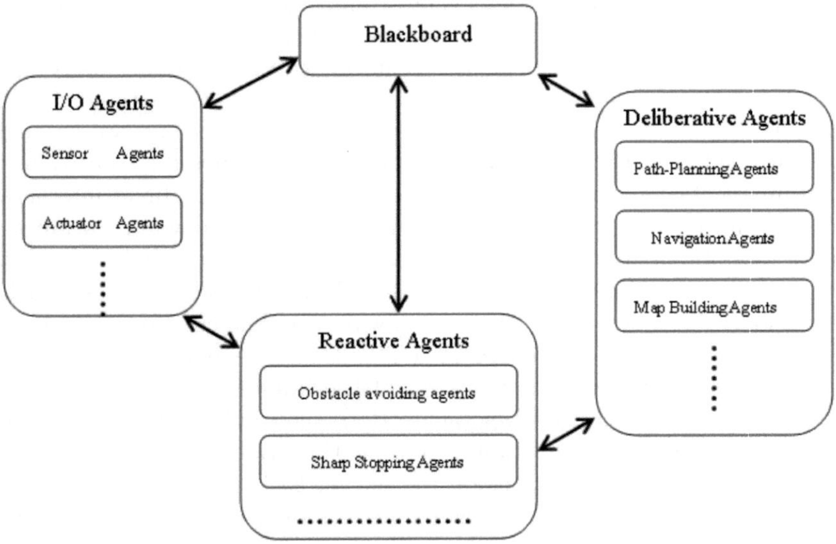


**Fig. 4.2.** Framework of multi-agent-based hybrid architecture

**I/O Agents**: Mostly refer to sensor agents and actuator agents. Sensor agents respond for detecting environment inside and outside, and after being preprocessed, sensor data is transferred into environment information and sent to blackboard. On the other hand, actuator agents receive commands and actuate the physical parts. Here, sensor agents not only include traditional sensors such as ultra-sonic sensor, CCD camera, infrared sensor, etc, but also include quantum sensors such as SQUID sensor and quantum well Hall sensor.

**Reactive Agents**: Map the sensor information to action, and send commands to actuator agents; Apart from I/O interface, reactive agents consist of two parts: (1) Library of reactive action rules; (2) Fusion of multi-action. Appropriate operating mechanism and high speed computing unit are necessary to guarantee the real time ability of reactive agents.

**Deliberate Agents**: Answer for task planning, such as reasoning, map-building, navigation, path-planning, etc. Their structures are multi-layered based on task decomposition and are specified according to different tasks and abstraction.

The hybrid architecture is a compromise between pure reactive and deliberative ones, which makes it much more complex and difficult to organize. But for the mobile system introduced above, the concept of multi-agent system is applied to the overall architecture, which makes it possible for the intelligent mobile system to acquire the following capabilities: (1) Adaptive learning; (2) Scalable ability; (3) Efficient and universal communication. The communication mechanism of MQAS provides versatile protocols for efficient and universal communication between different agents of the QIMS architecture.

The implementation of QIMS software architecture is divided into three levels: low level, middle level and high level. The high level software supplies an interface between human and system, which is mainly developed on the mainframe and is able to give command to mobile system platform when necessary. The middle level software is the collection of algorithms for different abstract tasks, such as navigation, localization, obstacle avoidance, etc. The low level software is developed to accomplish basic operations and also to execute decisions and commands given by the other two levels. In our research, we focus on the middle level, which is the most important and complex part of the software architecture for intelligent mobile systems. This software system has two factors: components and coordination. The components of the QIMS are based on multi-agent, so it is crucial to provide an effective method for coordination. Aiming at solving complex tasks, we use heterogeneous agents for different actions. The concept of interrupt is borrowed from computer science for switching between different agents according to different priority. For example, navigation is a very complex task which may consist of different sub-tasks such as path planning, wall-following, obstacle avoiding. These sub-tasks are generally accomplished by different kinds of agents. As shown in Fig.4.3, an obstacle avoidance action occurs when another action called wall-following is going on in QIMS navigation. As the priority of obstacle avoidance agent is higher than wall-following agent, the wall-following agent is de-activated and the obstacle avoidance agent is activated until the obstacle avoidance action is finished.

This kind of mechanism has at least three advantages: (1) coordinate different agents effectively, especially for agents with conflict; (2) make it easier to add new agents with certain functions; (3) compatible for algorithms programming, such as navigation algorithms based on hierarchical reinforcement learning [46].

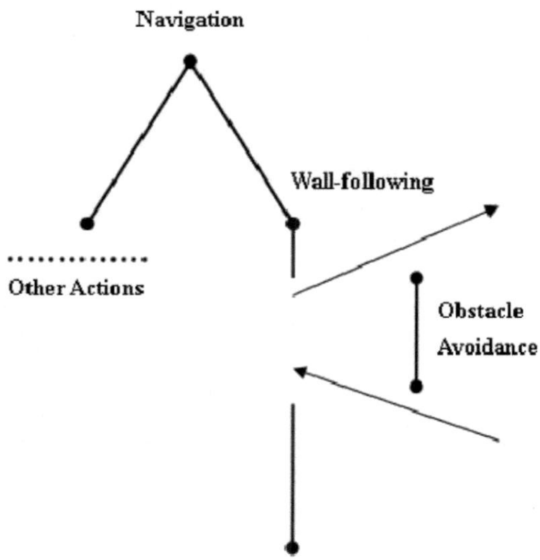


**Fig. 4.3.** An example of Coordination for agents

## 4.4 Hardware of typical QIMS

The overall hardware structure of the QIMS may be composed of two parts: quantum computers (or quantum information processing units, QIPU) and the classical-quantum hybrid components, such as multi-sensor systems, actuators and quantum or classical buses for communication.

### 4.4.1 Quantum computers

In the actual applications, quantum computers should be an important component of QIMS. A quantum computer is a physical apparatus that can process quantum information and perform parallel computation by manipulating quantum states in a controlled fashion. The most important advantage of quantum computer is quantum parallel computing ability. According to quantum computation theory, the quantum computing process can be looked upon as a unitary transformation $U$ from input qubits to output qubits. If one applies a transformation $U$ to a superposition state, the transformation will act on all basis vectors of this superposition state and the output will be a new superposition state by superposing the results of all basis vectors. So when one processes function $f(x)$ by the method, the transformation $U$ can simultaneously work out many different results for a certain input $x$. This is analogous with parallel process of classical computer, so it is called quantum parallelism. And the parallelism of quantum computation is one of the most important factors to acquire the powerful ability of quantum algorithm. Suppose the input qubit $|z\rangle$ lies in the superposition state:

$$|z\rangle = \alpha|0\rangle + \beta|1\rangle \tag{4.14}$$

The transformation $U_z$ which describes computing process may be defined as follows:

$$U_z : |z, y\rangle \rightarrow |z, y \oplus f(z)\rangle \tag{4.15}$$

where $|z, y\rangle$ represents the input joint state and $|z, y \oplus f(z)\rangle$ is the output joint state. $|y\rangle$ is an auxiliary qubit and '$\oplus$' means modulo 2 addition. Let $y = 0$ and we can easily obtain [1]:

$$U_z|z, y\rangle = \alpha|0, f(0)\rangle + \beta|1, f(1)\rangle \tag{4.16}$$

The result contains information about both $f(0)$ and $f(1)$, and we seem to evaluate $f(z)$ for two values of $z$ simultaneously.

The above process can correspond to a "quantum black box" (or oracle). By feeding quantum superposition states to a quantum black box, we can learn what is inside with an exponential speedup, compared to how long it would take if we were only allowed classical inputs [2]. Here, for input $|z\rangle = \alpha|0\rangle + \beta|1\rangle$, the quantum black box can compute both the values of $f(0)$ and $f(1)$ through running just once.

Since $U$ is a unitary transformation, computing function $f(x)$ can give [2]:

$$U \sum_{x=00...0}^{11...1} C_x|x, 0\rangle = \sum_{x=00...0}^{11...1} C_x U|x, 0\rangle = \sum_{x=00...0}^{11...1} C_x|x, f(x)\rangle \tag{4.17}$$

Based on the above analysis, it is easy to find that an $n$-qubit system can simultaneously process $2^n$ states. However, this is different from classical parallel computation, where multiple circuits built to compute are executed simultaneously, since quantum parallel computation doesn't necessarily make a tradeoff between computation time and needed physical space. In fact, quantum parallelism employs a single circuit to simultaneously evaluate the function for multiple values by exploiting the quantum state superposition principle and provides an exponential-scale computation space in the $n$-qubit linear physical space. Therefore quantum computation can effectively increase the computing speed of some important classical functions.

With the rapid development of quantum computation technology, some quantum computer models can be constructed using nuclear magnetic resonance (NMR), ion traps, and photons. In the present QIMS, quantum computers can accomplish some specific tasks such as storing, analyzing, computing, and processing a variety of information to help the QIMS conduct appropriate quantum control algorithms.

### 4.4.2 Classical-quantum hybrid components

From the whole hardware structure, we know that most of the QIMS is composed of classical and quantum ones.

For example, there are classical sensors and quantum sensors. We are familiar with the classical sensors, such as sonar sensors, CCD Cameras, infrared sensors, etc. But among the present QIMS, SQUID sensors and quantum well Hall sensors can also be used as special sensors so that the detecting system not only holds the common sensing ability, but also can measure super faint electromagnetic field, such as detecting cancer and tumor in biomedicine and sensing the faint signals of changing electromagnetic field in war industry.

Quantum Well Hall Sensor is a kind of high-performance micro Hall sensor and uses two-dimensional electron gases to obtain a compromise between high mobility and high carrier concentration while maintaining a reasonably high sheet resistance [47], [48]. For example, we can construct a quantum well Hall sensor through sandwiching thin InAs layers between AlGaSb layers and the sensor has high magnetic sensitivity and very excellent temperature stability as a result of a good confinement of two dimensional electron gases in quantum well structure. So quantum well Hall sensor can be used to detect faint electric magnetic field under different kinds of circumstances.

Superconduction quantum interference device (SQUID) sensor is extremely sensitive magnetic sensor that is based on the principles of superconductivity, the Meissner effect, flux quantisation and the Josephson effect [49]-[51]. As an essential part of a SQUID sensor, the Josephson junction is effectively a weak link between two superconductors capable of carrying supercurrents below a critical value. Using Josephson effect, SQUID sensor can convert minute changes in current or magnetic field to a measurable voltage and detect magnetic fields as small as $10^{-10}$ Tesla. So it has widely been applied as an ultra sensitive sensor for biomagnetism, nondestructive testing and environmental geophysics.

## 4.5 Planning for QIMS based on Grover algorithm

### 4.5.1 A general framework of planning tasks

Now let's consider the planning problem of QIMS whose state evolves according to certain transition probabilities that depend on a control $u$, if the QIMS is in state $i$ and chooses control $u$ according to a policy $\pi$, it will move to state $j$ with probability $p_{ij}(u)$ and a cost $g(i,u,j)$. Now suppose that the desirability of a state is defined as $V$, which means the value of a state. The task of QIMS planning is to find out the optimal sequence of $V(state)$, which satisfies some forms of Bellman's equation

$$V^*(i) = min_u E[g(i,u,j) + V^*(j)|i,u], for\ all\ i,j \tag{4.18}$$

where $E[\cdot|i,u]$ is the expected value. So at a certain state $i$ the QIMS planning problem is simply an unstructured searching problem to find the action $a_i$ (or the next state $j$) which is the optimal.

### 4.5.2 Grover algorithm

Grover algorithm is a quantum search algorithm proposed by Grover in 1997 [12]. If we have a very large unsorted database containing $N \gg 1$ items and we want to find one particular item as to find a needle in the haystack. Since the $N$ items are in a random order, we will need to look up $N/2$ items before the probability is 1/2 that we have found the wanted item. So the algorithm complexity is $O(N)$. However there is a quantum search algorithm (Grover algorithm) which enables this search method to be sped up substantially and the complexity reduces to $O(\sqrt{N})$.

### 4.5.3 Planning using Grover algorithm

In QIMS, the planning problem can be looked upon as an unstructured searching problem to find the optimal action $a$ from action space based on the current state. If the complexity of the state (or action) space is $O(N)$, the problem complexity for traditional intelligent mobile system is $O(N^2)$, whereas a QIMS can reduce the complexity to $O(N^{3/2})$ by using the Grover algorithm.

If there are $N$ alternative actions ($2^n \geq N \geq 2^{n-1}$), we can express them with $n$ qubits:

$$|\psi_0\rangle = \sum_{l=00\cdots0}^{11\cdots1} a_l|l\rangle \tag{4.19}$$

where the length of $l$ is $n$. For convenience, the fomula can also be expressed as

$$|\psi_0\rangle = \sum_{i=1}^{2^n} a_i|i\rangle \tag{4.20}$$

Assuming the action to be found corresponds to $|k\rangle$, now we use Grover algorithm to complete the search task.

At first, we prepare a quantum state

$$|s\rangle = \frac{1}{\sqrt{2^n}} \sum_{i=1}^{2^n} |i\rangle \tag{4.21}$$

which is an equal weight superposition state. This can be accomplished by applying the Hadamard transformation to each qubit of the $n$-qubit state $|00\cdots00\rangle$ [1]. Then we construct a reflection transform

$$U_s = 2|s\rangle\langle s| - I \tag{4.22}$$

Geometrically, when $U_s$ acts on an arbitrary vector, it preserves the component along $|s\rangle$ and flips the component in the hyperplane orthogonal to $|s\rangle$. Thus if we apply $U_s$ to $|\psi_0\rangle$, we get

$$U_s|\psi_0\rangle = 2|s\rangle\langle s|\psi_0\rangle - |\psi_0\rangle = \sum_{i=1}^{2^n}(2\langle a\rangle - a_i)|i\rangle \tag{4.23}$$

where

$$\langle a\rangle = \frac{1}{2^n}\sum_{i=1}^{2^n} a_i \tag{4.24}$$

Now we give another reflection transform

$$U_k = -2|k\rangle\langle k| + I \tag{4.25}$$

where $|k\rangle$ is the $k$-th eigenstate and by applying it to state $|\psi_0\rangle$, we obtain

$$U_k|\psi_0\rangle = -2|k\rangle\langle k|\psi_0\rangle + |\psi_0\rangle = \sum_{i=1,i\neq k}^{2^n} a_i|i\rangle - a_k|k\rangle \tag{4.26}$$

It is easy to see that $U_k$ only changes the amplitude's sign of $|k\rangle$ in the superposition state. Thus we can form a unitary transformation (Grover iteration) [12]

$$U_G = U_s U_k \tag{4.27}$$

By repeatedly applying the transformation $U_G$ on $|\psi_0\rangle$, we can enhance the probability amplitude of $|k\rangle$ while suppressing the amplitude of any other states $|i \neq k\rangle$ [1]. After enough number of iterations of the transformation, we can perform a measurement on the system to make the state $|\psi_0\rangle$ collapse into $|k\rangle$ with a probability of almost 1.

Define angle $\theta$ satisfying $\sin^2\theta = \frac{1}{2^n}$. After applying the $U_G$ $j$ times to $|\psi_0\rangle$, the amplitude of $|k\rangle$ becomes

$$a_k^j = \sin((2j+1)\theta) \tag{4.28}$$

Boyer has shown in [52] that the probability of failure is no more than $\frac{1}{2^n}$ if we perform the Grover iteration $int(\pi/4\theta)$ times (here the function $int(x)$ returns the integer part of $x$). According to Grover algorithm, when $N$ is large, QIMS can find the action corresponding to $|k\rangle$ with a high probability of $[1-O(1/N)]$. Since Grover algorithm can find desired result with a probability of almost 1 in $\sqrt{N}$ steps, the QIMS reduces the complexity and can faster find a suitable action from the action space based on the current state of the QIMS. If the number of states is $10^4$, considering different number of actions, the problem complexities in classical mobile systems and QIMS can be compared as shown in Table 4.1. From the Table, we can conclude that the advantage of QIMS is more and more prominent with the increase in the number of actions.

## 4.6 Learning for QIMS using QRL

The essence of QIMS learning is to deal with state-action pairs {State(t), Action(t)}. Here the widely used reinforcement learning (RL) methods [53]-[59]

**Table 4.1.** Comparisons of problem complexities in classical systems and QIMS

| Number of actions | $10^2$ | $10^3$ | $10^4$ | $10^5$ | $10^6$ | $10^7$ | $10^8$ |
|---|---|---|---|---|---|---|---|
| **classical systems** | $10^6$ | $10^7$ | $10^8$ | $10^9$ | $10^{10}$ | $10^{11}$ | $10^{12}$ |
| **QIMS** | $10^5$ | $3.2 \times 10^5$ | $10^6$ | $3.2 \times 10^6$ | $10^7$ | $3.2 \times 10^7$ | $10^8$ |

are introduced as an example to show how to improve traditional methods using quantum algorithm. Reinforcement learning uses a scalar value named reward to evaluate the input-output pairs and learns a mapping from states to actions by interaction with environment through trial-and-error. To adapt learning algorithms to a QIMS, we propose a novel learning method–quantum reinforcement learning (QRL), inspired by quantum superposition and quantum parallelism [40] [42].

### 4.6.1 QRL

If the numbers of states and actions are $N_s$ and $N_a$ respectively, we can choose numbers $m$ and $n$, which are characterized by the following inequalities:

$$N_s \leq 2^m \leq 2N_s, N_a \leq 2^n \leq 2N_a \tag{4.29}$$

And use $m$ and $n$ qubits to represent eigen state set $S = \{|s_i\rangle\}$ and eigen action set $A = \{|a_j\rangle\}$ respectively, we can obtain the corresponding relations as follows:

$$|s^{(N_s)}\rangle = \sum_{i=1}^{N_s} C_i |s_i\rangle \leftrightarrow |s^{(m)}\rangle = \sum_{s=00\cdots0}^{\overbrace{11\cdots1}^{m}} C_s |s\rangle \tag{4.30}$$

$$|a_{s_i}^{(N_a)}\rangle = \sum_{j=1}^{N_a} C_j |a_j\rangle \leftrightarrow |a_s^{(n)}\rangle = \sum_{a=00\cdots0}^{\overbrace{11\cdots1}^{n}} C_a |a\rangle \tag{4.31}$$

In other words, the states (or actions) of a QRL system may lie in the superposition state of eigen states (or eigen actions). Inequalities in (4.29) guarantee that every states and actions in traditional RL have corresponding representation with eigen states and eigen actions in QRL. The probability amplitude $C_s$ and $C_a$ are complex numbers and satisfy

$$\sum_{s=00\cdots0}^{\overbrace{11\cdots1}^{m}} |C_s|^2 = 1 \tag{4.32}$$

$$\sum_{a=00\cdots0}^{\overbrace{11\cdots1}^{n}} |C_a|^2 = 1 \tag{4.33}$$

since $|C_s|^2$ and $|C_a|^2$ are the probabilities with which we get eigen state $s$ or eigen action $a$ when we measure the related quantum state or action, the probabilities must sum to one.

The procedural form of a standard QRL algorithm can be described as Algorithm 1. In QRL algorithm, after initializing the state and action we can observe $|a_s^{(n)}\rangle$ and obtain an eigen action $|a\rangle$. Excute this action and the system can give out next state $|s'\rangle$, reward $r$ and state value $V(s')$. $V(s)$ is updated by TD(0) rule, and $r$ and $V(s)$ can be used to determine the Grover iteration times. The Grover iteration $U_{Grov}$ is constructed by two reflections $U_{a_0^{(n)}}$ and $U_a$, where $|a_0^{(n)}\rangle$ is equally weighted superposition of all computational basis actions. $U_a$ flips the sign of the action $|a\rangle$, but acts trivially on any action orthogonal to $|a\rangle$. On the other hand, $U_{a_0^{(n)}}$ preserves $|a_0^{(n)}\rangle$, but flips the sign of any vector orthogonal to $|a_0^{(n)}\rangle$.

***Algorithm 1: Quantum reinforcement learning (QRL)***

1. initialize $|s^{(m)}\rangle = \sum_{s=00\cdots0}^{\overbrace{11\cdots1}^{m}} C_s|s\rangle$, $f(s) = |a_s^{(n)}\rangle = \sum_{s=00\cdots0}^{\overbrace{11\cdots1}^{m}} C_a|a\rangle$ and $V(s)$ arbitrarily;
2. repeat (for each episode)
3. for all states $|s\rangle$ in $|s^{(m)}\rangle = \sum_{s=00\cdots0}^{\overbrace{11\cdots1}^{m}} C_s|s\rangle$:
4. observe $f(s) = |a_s^{(n)}\rangle$ and get $|a\rangle$;
5. take action $|a\rangle$, observe next state $|s^{'}\rangle$, reward $r$, then
6. (a) update state value: $V(s) \leftarrow V(s) + \alpha(r + \gamma V(s^{'}) - V(s))$
7. (b) update probability amplitudes:
8. repeat for $[k(r + V(s^{'}))]$ times:
9. $U_{Grov}|a_s^{(n)}\rangle = U_{a_0^{(n)}} U_a |a_s^{(n)}\rangle$
10. until for all states $|\Delta V(s)| \leq \varepsilon$

end.

QRL is inspired by the superposition principle of quantum state and quantum parallel computation. The action set can be represented with quantum state and the eigen action can be obtained by randomly observing the simulated quantum state, which will lead to state collapse according to quantum measurement postulate. And the occurrence probability of eigen action is determined by corresponding probability amplitude, which is updated according to rewards and value functions. So this approach represents the whole state-action space with the superposition of quantum state and makes a good tradeoff between exploration and exploitation using probability.

The merit of QRL is dual. First, as for simulation algorithm on traditional computer it is an effective algorithm with novel representation and computation methods. Second, the representation and computation mode are consistent with quantum parallel computation system and can speed up learning with quantum computer or quantum gates.

### 4.6.2 Learning in unknown environment using QRL

Now let's consider a typical gridworld example for QIMS learning. The gridworld environment is as shown in Fig.4.4 and each cell of the grid corresponds to an individual state (eigen state) of the environments. From any state the agent can perform one of four primary actions (eigen actions): up, down, left and right, and actions that would lead into a blocked cell are not executed. The task of the algorithms is to find an optimal policy which will let the QIMS (agent) move from start point $S$ to goal point $G$ with minimized cost (or maximized rewards).

#### Experimental set-up

In this $13 \times 13$ $(0 \sim 12)$ grid world, the initial state $S$ and the goal $G$ is cell(4,4) and cell(8,8) and before learning the agent has no information about the environment at all. Once the agent finds the goal state it receives a reward of 100 and then ends this episode. All steps are punished by a reward of -1. The discount factor $\gamma$ is set to 0.99 for all the algorithms that we have carried out. In the first experiment, we compare QRL algorithm with TD(0) and we also demonstrate the expected result on a quantum computer theoretically. For the action selection policy of TD algorithm, we use $\epsilon$-greedy policy ($\epsilon = 0.01$). As for QRL, the action selecting policy is obviously different from traditional RL algorithms, which is inspired by the collapse theory of quantum measurement. And the value of $|C_a|^2$ is used to denote the probability of an action defined as $f(s) = |a_s^{(n)}\rangle = \sum_{a=00\cdots0}^{11\cdots1} C_a|a\rangle$. For the four cell-to-cell actions $|C_a|^2$ is initialized uniformly.

#### Experimental results and analysis

Learning performance for QRL compared with TD algorithm is plotted in Fig.4.5. We observe that QRL is also an effective algorithm on traditional computer although it is inspired by the quantum mechanical system and is designed for quantum computer in the future. As shown in Fig.4.5, QRL explores more than TD at the beginning of learning phase, but it learns much faster and guarantees a better balancing between exploration and exploitation. In addition, it is much easier to tune the parameters for QRL algorithms

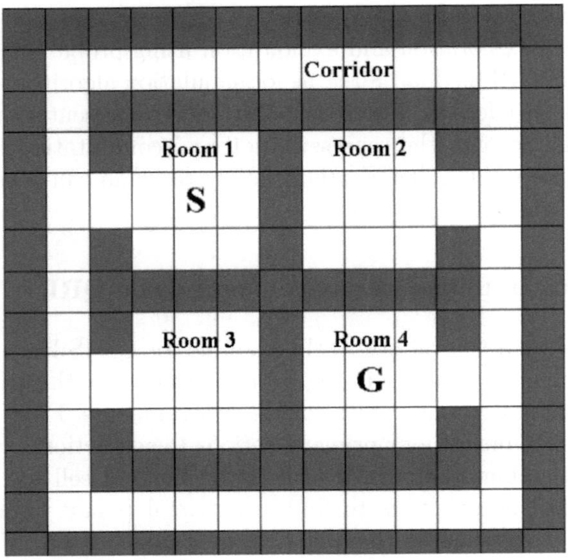


**Fig. 4.4.** The gridworld environment with cell-to-cell actions (up, down, left and right).

than for traditional ones. What's more important, according to the theoretical results, QRL has great potential of powerful computation provided that quantum computer (or related quantum apparatuses) is available in the future, which will lead to a more effective approach for the existing problems of learning in complex environments.

Furthermore, Fig.4.6 illustrates the results of QRL algorithms with different learning rates: $\alpha$ (alpha), from 0.02 to 0.20. From these figures, it can been concluded that given a proper learning rate (alpha<0.10) this algorithm learns fast and explores much at the beginning phase, and then steadily converges to the optimal policy that costs 25 steps to the goal $G$. As the learning rate increases from 0.02 to 0.10, this algorithm learns faster but more unsteadily. When the learning rate is 0.20 or above, it cannot converge to the optimal policy.

All the results show that QRL algorithm for the learning of QIMS is effective and excels other RL algorithms in the following two main aspects: (1) Action selecting policy makes a good tradeoff between exploration and exploitation using probability, which speeds up the learning and guarantees the searching over the whole state-action space as well. (2) Representation is based on the superposition principle of quantum mechanics and the updating process is carried out through parallel computation, which will be much more prominent in the future when practical quantum apparatus comes into use instead of being simulated on traditional computers.

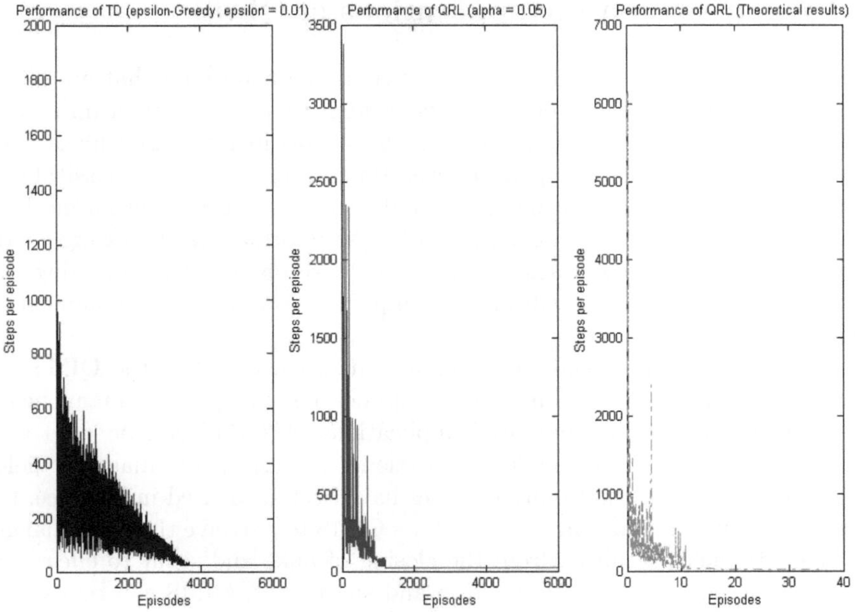


**Fig. 4.5.** Performance of QRL compared with TD algorithm ($\epsilon$-greedy policy).

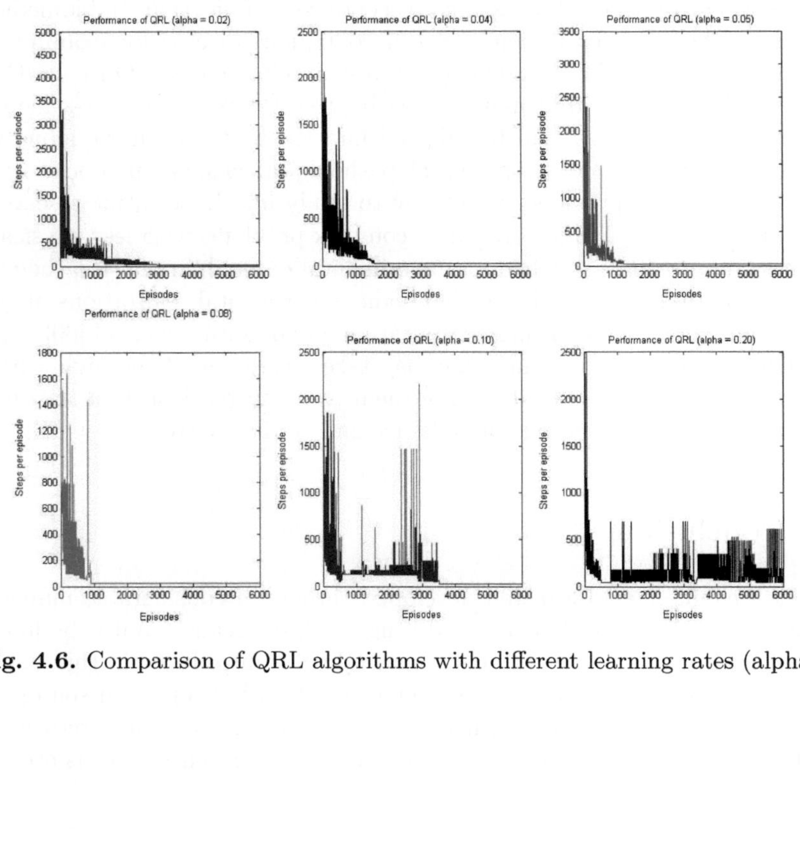


**Fig. 4.6.** Comparison of QRL algorithms with different learning rates (alpha).

## 4.7 Discussion on the applications of QIMS

Now we can see that QIMS has many obvious characteristics that are different from those of classical one. (1) The system property is quantum system or quantum-classical hybrid system; (2) The sensors include quantum sensors, such as SQUID sensors and quantum well Hall sensors, as well as classical sensors; (3) The physical laws that we should keep to include quantum mechanical law and classical mechanical law; (4) The information processing centre may run quantum computation, which may take on parallel processing; (5) Quantum communication will play an important role in the communication manner.

Due to so many special characteristics mentioned above, the QIMS has many advantages and will have wide application prospect in many fields. Here, we outline several potential applications. A QIMS can be used as a patrol warrior in military affairs and national defense. If the magnetic fields near important ports and military bases have been measured in advance, the high sensitivity of quantum sensors allows QIMS to perceive the faint changes in magnetic fields resulting from the closing of nuke and scout to effectively forewarn decision-makers. In aviation and spaceflight, QIMS can be used as Mars and lunar exploration vehicles with its advantage of high sensitivity in perceiving environments, its powerful ability to process information and its inherence nature that allows for more secure communication. In biomedicine, QIMS can be used for patients with a contagious disease for examining or tracking state of an illness, and can act as contagion doctors. Once a QIMS is successfully constructed, it may be possible to establish molecular-scale medical robots. Thus QIMS can be injected into the body and move along with blood circulation to detect potential pathological changes in body. At the same time, it may provide a novel way to study life. In scientific research, it is possible to use QIMS to solve most complex problems with less physical resources and provide a test-bed for experimental research in physics, chemistry and information technologies so that some experimental realizations of quantum communication, quantum computation and quantum control [60] may be possible. In safety engineering, one may take advantage of the high sensitivity and small size of QIMS to apply them to many tasks such as anti-terror forewarning, fire forecasting, surveillance and traffic control.

## 4.8 Conclusions

Mobile systems have been well developed through a considerable number of theoretical and practical advances during the last decades. With the foundation of traditional mobile systems, a new QIMS is introduced in this chapter. The QIMS takes most advantage of high sensitivity of quantum sensors and parallel ability of quantum algorithms. In order to manage these heterogeneous modules of the QIMS, a hybrid architecture based on multi-agent is proposed to harmonize different modules. To detect super faint signals, a multi-sensor system is designed based on SQUID sensor and quantum well Hall sensor. Grover algorithm is applied to perform planning task with complexity $O(N^2)$ and the theoretic result shows that QIMS can reduce the problem complexity to $O(N^{3/2})$. Then fusing quantum parallel algorithm in traditional reinforcement learning method, we propose a novel quantum reinforcement learning method for the QIMS and give a simple simulation example to demonstrate the validity and superiority of QRL. Compared with traditional mobile system, the QIMS effectively overcomes the limitation of existing sensors' performance and is able to meet real-time requirement for complex problems, which will make it competent for applications in various fields such as military affairs, aviation, biomedicine, safety engineering, etc. What's more, following this research, applicable real quantum robot [32] will also be designed and implemented, which will lead to new approaches for solving most complex problems with less physical resources, where the computational complexity is very high, the environment is full of uncertainty, or the quantum effect can not be neglect.

Although the research work introduced in this chapter is only the first step towards practical QIMS, it has a promising future given the rapid development and gradual maturation of quantum information technology and quantum control theory. To implement a real QIMS is no doubt a very challenging mission, which consists of four kinds of work: (1) synthesis architecture for QIMS, including classical-quantum hybrid systems; (2) physical implementation, including computing units, sensors, actuators and communication hardware; (3) software level researches, such as related theories and programming issues, including quantum-inspired algorithm, quantum algorithm, quantum simulation, algorithm integration, etc.

## References


1. Nielsen MA, Chuang IL (2000) Quantum Computation and Quantum Information. Cambridge, England, Cambridge University Press
2. Preskill J (1998) Physics 229: Advanced Mathematical Methods of Physics–Quantum Information and Computation. California Institute of Technology. Available electronically via http://www.theory.caltech.edu/people/preskill/ph229/
3. Wootters WK, Zurek WH (1982) A single quantum cannot be cloned. Nature, 299: 802-803
4. Benioff P (1980) The computer as a physical system: A microscopic quantum mechanical Hamiltonian model of computers as represented by Turing machines. Journal of Statistical Physics, 22: 563-591
5. Feynman RP (1982) Simulating physics with computers. International Journal of Theoretical Physics, 21: 467-488
6. Feynman RP (1986) Quantum mechanical computers. Foundations of Physics, 16: 507-531

7. Deutsch D (1985) Quantum theory, the Church-Turing principle and the universal quantum computer. Proceedings of The Royal Society of London Series A, 400: 97-117
8. Bennett CH, Brassard G (1984) Quantum cryptography: Public key distribution and coin tossing. Proceedings of the IEEE International Conference on Computers, Systems and Signal Processing, pp.175-179, IEEE Press, New York
9. Shor PW (1994) Algorithms for quantum computation: discrete logarithms and factoring. in Proceedings of the 35th Annual Symposium on Foundations of Computer Science, pp.124-134, IEEE Press, Los Alamitos, CA
10. Ekert A, Jozsa R (1996) Quantum computation and Shor's factoring algorithm. Reviews of Modern Physics, 68: 733-753
11. Grover LK (1996) A fast quantum mechanical algorithm for database search. in Proceedings of the 28th Annual ACM Symposium on the Theory of Computation, pp.212-219, ACM Press, New York
12. Grover LK (1997) Quantum mechanics helps in searching for a needle in a haystack. Physical Review Letters, 79: 325-327
13. Vandersypen LMK, Steffen M, Breyta G, Yannoni CS, Sherwood MH, Chuang IL (2001) Experimental realization of Shor's quantum factoring algorithm using nuclear magnetic resonance. Nature, 414: 883-887
14. Chuang IL, Gershenfeld N, Kubinec M (1998) Experimental implementation of fast quantum searching. Physical Review Letters, 80: 3408-3411
15. Jones JA (1998) Fast searches with nuclear magnetic resonance computers. Science, 280: 229-229
16. Jones JA, Mosca M, Hansen RH (1998) Implementation of a quantum Search algorithm on a quantum computer. Nature, 393: 344-346
17. Kwiat PG, Mitchell JR, Schwindt PDD, White AG (200) Grover's search algorithm: an optical approach. Journal of Modern Optics, 47: 257-266
18. Scully MO, Zubairy MS (2001) Quantum optical implementation of Grover's algorithm. Proceedings of the National Academy of Sciences of the United States of America, 98: 9490-9493
19. Meyer DA (1999) Quantum strategies. Physical Review Letters, 82: 1052-1055
20. Eisert J, Wilkens M, Lewenstein M (1999) Quantum games and quantum strategies. Physical Review Letters, 83: 3077-3080
21. Du JF, Li H, Xu XD, Shi MJ, Wu JH, Zhou XY, Han RD (2002) Experimental realization of quantum games on a quantum computer. Physical Review Letters, 88: 137902
22. Vala J, Amitay Z, Zhang B, Leone SR, Kosloff R (2002) Experimental implementation of the Deutsch-Jozsa algorithm for three-qubit functions using pure coherent molecular superpositions. Physical Review A, 66: 062316
23. Bennett CH, Brassard G, Crépeau C, Jozsa R, Peres A, Wootters W (1993) Teleporting an unknown quantum state via dual classical and Einstein-Podolsky-Rosen channels. Physical Review Letters, 70: 1895-1899
24. Bouwmeester D, Pan JW, Mattle K, Eibl M, Weinfurter H, Zeilinger A (1997) Experimental quantum teleportation. Nature, 390: 575-579
25. Furusawa A, Srensen JL, Braunstein SL, Fuchs CA, Kimble HJ, Polzik ES (1998) Unconditional quantum teleportation. Science, 282: 706-709
26. Nielsen MA, Knill E, Laflamme R (1998) Complete quantum teleportation using nuclear magnetic resonance. Nature, 396: 52-55

27. Pan JW, Bouwmeester D, Weinfurter H, Zeilinger A (1998) Experimental entanglement swapping: Entangling photons that never interacted. Physical Review Letters, 80: 3891-3894
28. Pan JW, Gasparoni S, Ursin R, Weihs G, Zeilinger A (2003) Experimental entanglement purification of arbitrary unknown states. Nature, 423: 417-422
29. Zhao Z, Chen YA, Zhang AN, Yang T, Briegel HJ, Pan JW (2004) Experimental demonstration of five-photon entanglement and open-destination teleportation. Nature, 430: 54-58
30. Lu CY, Zhou XQ, ühne OG, Gao WB, Zhang J, Yuan ZS, Goebel A, Yang T, Pan JW (2007) Experimental entanglement of six photons in graph states. Nature Physics, doi:10.1038/nphys507, Published online: 14 January 2007
31. Benioff P (1998) Quantum Robots and Environments. Physical Review A, 58: 893-904
32. Dong DY, Chen CL, Zhang CB, Chen ZH (2006) Quantum robot: structure, algorithms and applications. Robotica, 24: 513-521
33. Kak S (1995) On quantum neural computing. Information Sciences, 83: 143-160
34. Ventura D, Martinez T (2000) Quantum associative memory. Information Sciences, 124: 273-296
35. Narayanan A, Menneer T (2000) Quantum artificial neural network architectures and components. Information Sciences, 128: 231-255
36. Behrman EC, Nash LR, Steck JE, Chandrashekar VG, Skinner SR (2000) Simulations of quantum neural networks. Information Sciences, 128: 257-269
37. Rigatos GG, Tzafestas SG (2002) Parallelization of a fuzzy control algorithm using quantum computation. IEEE Transactions on Fuzzy Systems, 10(4): 451-460
38. Venayagamoorthy GK, Singhal G (2005) Quantum-inspired evolutionary algorithms and binary particle swarm optimization for training MLP and SRN neural networks. Journal of Computational and Theoretical Nanoscience, 2(4): 561-568
39. Sahin M, Atav U, Tomak M (2005) Quantum genetic algorithm method in self-consistent electronic structure calculations of a quantum dot with many electrons. International Journal of Modern Physics C, 16(9): 1379-1393
40. Dong DY, Chen CL, Chen ZH (2005) Quantum reinforcement learning. in Proceedings of First International Conference on Natural Computation, Lecture Notes in Computer Science, 3611: 686-689
41. Dong DY, Chen CL, Chen ZH, Zhang CB (2006) Quantum mechanics helps in learning for more intelligent robots. Chinese Physics Letters, 23: 1691-1694
42. Chen CL, Dong DY, Chen ZH (2006) Quantum computation for action selection using reinforcement learning. International Journal of Quantum Information, 4: 1071-1083
43. Brennen GK, Song D, Williams CJ (2003) Quantum-Computer Architecture Using Nonlocal Interactions. Physical Review A, 67: 050302
44. Brooks RA (1986) A robust layered control system for a mobile robot. IEEE Journal of Robotics and Automation, 2: 14-23
45. Barberá HM (2001) A distributed architecture for intelligent control in autonomous mobile robots. Doctor thesis, University of Murcia, Spain
46. Barto AG, Mahanevan S (2003) Recent advances in hierarchical reinforcement learning. Discrete Event Dynamic Systems: Theory and Applications, 13: 41-77
47. Behet M, Bekaert J, De Boeck J, Borghs G (2000) InAs/Al0.2Ga0.8Sb quantum well Hall effect sensors. Sensors and Actuators, 81: 13-17

48. Behet M, Das J, De Boeck J, Borghs G (1998) InAs/(Al, Ga)Sb quantum well structure for magnetic sensors. IEEE Transactions on Magnetics, 34: 1300-1302
49. Kallias G, Devlin E, Christides C, Niarchos D (2002) High $T_c$ SQUID sensor system for nondestructive evaluation. Sensors and Actuators, 85: 239-243
50. Mahdi AE, Mapps DJ (2000) High-$T_c$ SQUIDs: the ultra sensitive sensors for nondestructive testing and biomagnetism. Sensors and Actuators, 81: 367-370
51. Morisawa J, Otaka M, Kodama M, Kato T, Suzuki S (2002) Detection of intergranular cracking susceptibility due to hydrogen in irradiated austenitic stainless steel with s superconducting quantum interference device (SQUID) sensor. Journal of Nuclear Materials, 302: 66-71
52. Boyer M, Brassard G, Høyer P (1998) Tight bounds on quantum searching. Fortschritte Der Physik-Progress of Physics, 46: 493-506
53. Sutton R, Barto AG (1998) Reinforcement Learning: An Introduction. Cambridge, MA: MIT Press
54. Kaelbling LP, Littman ML, Moore AW (1996) Reinforcement learning: A survey. Journal of Artificial Intelligence Research, 4: 237-287
55. Sutton R (1988) Learning to predict by the methods of temporal difference. Machine Learning, 3: 9-44
56. Watkins C, Dayan P (1992) Q-learning. Machine Learning, 8: 279-292
57. Kaya M, Alhajj A (2005) A novel approach to multiagent reinforcement learning: Utilizing OLAP mining in the learning process. IEEE Transactions on Systems Man and Cybernetics C, 35: 582-590
58. Even-Dar E, Mansour Y (2003) Learning rates for Q-learning. Journal of Machine Learning Research, 5: 1-25
59. Whiteson S, Stone P (2006) Evolutionary function approximation for reinforcement learning. Journal of Machine Learning Research, 7: 877-917
60. Chen ZH, Dong DY, Zhang CB (2005) Quantum Control Theory: An Introduction. Hefei, China: University of Science and Technology of China Press (in Chinese)

# 5

# Quantum Associative Pattern Retrieval

Carlo A. Trugenberger[1] and Cristina M. Diamantini[2]

[1] SwissScientific.com, ch. Diodati 10, CH-1223 Cologny
ca.trugenberger@bluewin.ch
[2] Dipartimento di Fisica, Universitá di Perugia, via A. Pascoli, I-06100 Perugia, Italy cristina.diamantini@pg.infn.it

Associative pattern retrieval, one of the hallmarks of intelligence, cannot only be realized by the traditional attractor dynamics of the Hopfield model but also by a reversible, unitary evolution of quantum bits (qubits). We will show that qubit networks with long-range interactions governed by the Hebb rule can be used as quantum associative memories. Starting from a uniform superposition, the unitary evolution generated by these interactions drives the network through a quantum phase transition at a critical computation time, after which ferromagnetic order guarantees that a measurement retrieves the stored patterns. The memory capacity of these qubit networks depends on the computation time: the maximum capacity is reached at a memory density $\alpha = p/n = 1$, after which a phase transition to a quantum spin glass state implies total amnesia. At these loading factors, however the retrieval quality is poor; admitting only a few percent of errors requires lower memory loading factors, comparable with the classical Hopfield model.

## 5.1 Introduction

Historically, the interest in neural networks [1] has been driven by the desire to build machines capable of performing tasks for which the sequential circuit paradigm of Babbage and von Neumann are not well suited, like pattern recognition, categorization and generalization. Since these higher cognitive tasks are typical of biological intelligences, the design of these parallel distributed processing systems has been largely inspired by the physiology of the human brain.

On the other side, the last decade has seen the emergence of quantum mechanics as a powerful new paradigm for computation [2]. To a large extent these development have focused only on the quantum circuit model, in which a set of elementary universal quantum gates are sequentially applied on qubit registers. The root of the interest in quantum computation lies mainly in

C.A. Trugenberger and C.M. Diamantini: *Quantum Associative Pattern Retrieval*, Studies in Computational Intelligence (SCI) **121**, 103–113 (2008)
www.springerlink.com 

the speed-up in computing time it can provide with respect to its classical counterpart.

The natural question arises, if a quantum version of neural networks can be formulated. We will show below, on the concrete example of the Hopfield model [3], that this question can be answered in the affirmative. We shall in fact describe a new quantum information processing paradigm which is capable of higher cognitive tasks as described above [4]. On the other hand, the model we shall present can also be viewed as the quantization of the Hopfield model, thereby closing the circle. Why the Hopfield model? Because the complete feed-back loops between neurons in this model lead to associative pattern recognition and this capability is probably one of the basic hallmarks of intelligence.

Instead of one- and two-qubit interactions that are switched on and off sequentially, we will consider fixed, long-range interactions inspired by neural networks. The resulting quantum system is a fully-connected interacting network of qubits in which memories are encoded in the long-range interaction matrix: the corresponding Hamiltonian generates a unitary evolution capable of memory retrieval and pattern recognition. Note the crucial difference with respect to the classical Hopfield model; this is governed by an irreversible attractor dynamics, its quantum counterpart, instead, retrieves patterns by a reversible, unitary dynamics.

Similar interacting qubit networks have been considered recently for quantum state transfer [5]. Note also that some of the most promising technologies for the implementation of quantum information processing, like optical lattices [6] and arrays of quantum dots [7] rely exactly on similar collective phenomena.

## 5.2 Quantizing Neural Networks

In mathematical terms, the simplest neural network model is a directed graph with the following properties:

1. A state variable $n_i$ is associated with each node (neuron) $i$.
2. A real-valued weight $w_{ij}$ is associated with each link (synapse) $(ij)$ between two nodes $i$ and $j$.
3. A state-space-valued transfer function $f(h_i)$ of the synaptic potential $h_i = \sum_j w_{ij} n_j$ determines the dynamics of the network.

Two types of dynamical evolution have been considered: sequential or parallel synchronous. In the first case the neurons are updated one at a time according to

$$n_i(t+1) = f\left(\sum_k w_{ik} n_k(t)\right) , \tag{5.1}$$

while in the second case all neurons are updated at the same time. The simplest model is obtained when neurons become binary variables taking only the values $n_i = \pm 1$ for all $i$ and the transfer function, called activation function in this case, becomes the sign function, $f(h) = \text{sign}(h)$. This is the original McCullogh-Pitts [8] neural network model, in which the two states represent quiescent and firing neurons.

The Hopfield model [3] is a fully-connected (each neuron is coupled to all others) McCullogh-Pitts network in which the synaptic weights are symmetric quantities chosen according to the Hebb rule [1]

$$w_{ij} = w_{ji} = \frac{1}{n-1} \sum_{\mu=1}^{p} \xi_i^{\mu} \xi_j^{\mu} , \qquad w_{ii} = 0 . \tag{5.2}$$

Here, $n$ is the total number of neurons and $\xi^{\mu}$ are $p$ binary patterns to be memorized ($\xi_i^{\mu} = \pm 1$).

For $p/n < 0.138$ the network works as an associative (or content-addressable) memory which permits recall of information only on the basis of partial knowledge of its content, without any further information on storage location. Given an initial input, the network configuration will converge to the stored pattern closest to it in Hamming distance (the number of different bits). The stored patterns are thus attractors for the dynamics (5.1) defined by the transfer function. For higher loading factors $p/n$, however, there is a phase transition to a spin glass phase [9] which impairs the ability to correctly recall memories.

The "quantization" of Hopfield networks can be carried out by substituting each neuron with a qubit i.e. a quantum degree of freedom with a two-dimensional Hilbert space whose basis states can be labeled as $|0>$ and $|1>$. Classical state variables at each node are substituted by the third Pauli matrix $\sigma^z$ acting on the corresponding Hilbert space. This has the eigenvalues $\pm 1$ that correspond to quiescent and firing neurons: note however, that, contrary to the classical model, a neuron can be in a linear superposition of its quiescent and firing state before a measurements projects it onto one of the basis states.

The only tricky point regards the treatment of interactions bewteen the qubits. In the classical model the dynamics (5.1) induced by the transfer function is fully deterministic and irreversible, which is not compatible with quantum mechanics. A first generalization that has been considered is that of stochastic neurons, in which the transfer function determines only the probabilities that the classical state variables will take one of the two values: $n_i(t+1) = \pm 1$ with probabilities $f(\pm h_i(t))$, where $f$ must satisfy $f(h \to -\infty) = 0$, $f(h \to +\infty) = 1$ and $f(h) + f(-h) = 1$.

While this modification makes the dynamics probabilistic by introducing thermal noise, the evolution of the network is still irreversible since the actual *values* of the neurons are prescribed after an update step. In quantum mechanics the evolution must be reversible and only the magnitudes of the *changes* in the neuron variables can be postulated. Actually, the dynamics must generate a *unitary* evolution of the network, the simplest example consisting of a product of rotations in the Hilbert spaces of each qubit.

In [4] we proposed the following model. The interactions between the qubits of the network are governed by the following "transverse" Hamiltonian:

$$\mathcal{H} = J \sum_{ij} w_{ij} \sigma_i^y \sigma_j^z \ , \tag{5.3}$$

where $\sigma^k$, $k = x, y, z$ denote the Pauli matrices and $J$ is a coupling constant with the dimensions of mass (we use units $c = 1, \hbar = 1$). As we now show, the synaptic potential $h_i = \sum_j w_{ij} \sigma_j^z$, aggregating the quantum states of all other qubits, generates a transverse magnetic field along the $y$-axis which rotates qubit $i$ towards the $z$-axis, along which it is then measured.

The Hamiltonian (5.3) generates a unitary evolution of the network:

$$|\psi(t)> = \exp(i\mathcal{H}t) \ |\psi_0> \ , \tag{5.4}$$

where $|\psi_0> = |\psi(t=0)>$. Specifically, we will choose as initial configuration of the network the uniform superposition of all computational basis states,

$$|\psi_0> = \frac{1}{\sqrt{2^n}} \sum_{x=0}^{2^n-1} |x> \ , \tag{5.5}$$

where we have used the standard notation in which $|x>$ denotes the state determined by the integer $x$ in binary format. This corresponds to a "blank memory" in the sense that all possible states have the same probability of being recovered upon measurement. In the language of spin systems this is a state in which all spins are aligned in the $x$ direction, corresponding to the maximal rate of quantum tunneling bewteen the up and down $\sigma_3$ states.

Inputs $\xi^{\rm ext}$ can be accomodated by adding an external transverse magnetic field along the $y$ axis, i.e. modifying the Hamiltonian to

$$\begin{aligned} \mathcal{H} &= J \sum_{ij} w_{ij} \sigma_i^y \sigma_j^z + g \sum_i h_i^{\rm ext} \sigma_i^y \ , \\ h_i^{\rm ext} &= \sum_j w_{ij} \xi_j^{\rm ext} \ . \end{aligned} \tag{5.6}$$

This external magnetic field can be thought of as arising from the interaction of the network with an additional "sensory" qubit register prepared in the state $\xi^{\rm ext}$, the synaptic weights between the two layers being identical to those of the network self-couplings. The dimensionless quantity $g/J$ determines the relative strength of the external magnetic field with respect to the internal dynamics.

## 5.3 Quantum Associative Memories

Let us now specialize to the simplest case of one assigned memory $\xi$ in which $w_{ij} = \xi_i \xi_j$. In the classical Hopfield model there are two nominal stable states that represent attractors for the dynamics, the pattern $\xi$ itself and its negative $-\xi$: all initial configurations that deviate from the assigned memory by less than half the bits converge towards $\xi$ after one global update while the other configurations converge towards its negative $-\xi$. Thus all initial configurations end up in nominal assigned memories. Correspondingly, the quantum dynamics defined by the Hamiltonian (5.3) and the initial state (5.5) have a $Z_2$ symmetry generated by $\prod_i \sigma_i^x$, corresponding to the inversion $|0> \leftrightarrow |1>$ of all qubits.

Since the qubits of the network are fully connected and weakly interacting, the model can be expected to be exactly solvable in the $n \to \infty$ limit by a mean-field theory. Indeed, this is known to become exact for weak, long-range interactions [10]. In the mean-field approximation operators are decomposed in a sum of their mean values and fluctuations around it, $\sigma_i^k = < \sigma_i^k > + (\sigma_i^k - < \sigma_i^k >)$ and quadratic terms in the fluctuations are neglected in the Hamiltonian. Apart from an irrelevant constant, this gives

$$\begin{aligned} \mathcal{H}_{\text{mf}} &= J \sum_i \sigma_i^y \left( < h_i^z > + \frac{g}{J} h_i^{\text{ext}} \right) + \sigma_i^z < h_i^y > , \\ < h_i^k > &= \sum_j w_{ij} < \sigma_j^k > = \xi_i \, m^k , \end{aligned} \tag{5.7}$$

where $m^k = (1/n) \sum_i < \sigma_i^k > \xi_i$ is the average overlap of the state of the network with the stored pattern. This means that each qubit $i$ interacts with the average magnetic field (synaptic potential) $< h_i^k >$ due to all other qubits: naturally, the correct values of these mean magnetic fields $< h_i^k >$ have to be determined self-consistently.

To this end we compute the average pattern overlaps $m^k$ using the mean field Hamiltonian (5.7) to generate the time evolution of the quantum state. This reduces to a sequence of factorized rotations in the Hilbert spaces of each qubit, giving

$$\begin{aligned} m^y &= -\frac{m^y}{|m|} \sin 2Jt|m| , \\ m^z &= \frac{m^z + (g/J)M^z}{|m|} \sin 2Jt|m| , \end{aligned} \tag{5.8}$$

where $|m| = \sqrt{(m^y)^2 + (m^z + (g/J)M^z)^2}$ and $M^z = (1/n) \sum_i \xi_i^{\text{ext}} \xi_i$ is the average overlap of the external stimulus with the stored memory. Note that, even for $g = 0$, the symmetry of the Hamiltonian under exchange of the y and z axes is broken by the initial state with all qubit spins aligned along the x axe.

Before we present the detailed solution of these equations, let us illustrate the mechanism underlying the quantum associative memory. To this end we note that, for $g = 0$, the pattern overlaps $m^y$ and $m^z$ in the two directions cannot be simultaneously different from zero. As we show below, only $m^z \neq 0$

for $J > 0$ (for $J < 0$ the roles of $m^y$ and $m^z$ are interchanged). In this case the evolution of the network becomes a sequence of $n$ rotations

$$\begin{pmatrix} \cos(Jt < h_i^z >) & \sin(Jt < h_i^z >) \\ -\sin(Jt < h_i^z >) & \cos(Jt < h_i^z >) \end{pmatrix} \tag{5.9}$$

in the two-dimensional Hilbert spaces of each qubit $i$. The rotation parameter is exactly the same synaptic potential $h_i$ which governs the classical dynamics of the Hopfield model. When these rotations are applied on the initial state (5.5) with all qubits $(1/\sqrt{2}\ 1/\sqrt{2})^T$ aligned along the $x$ axis, they amount to an update step transforming the qubit spinors into

$$\frac{1}{\sqrt{2}} \begin{pmatrix} \cos(Jt < h_i^z >) + \sin(Jt < h_i^z >) \\ \cos(Jt < h_i^z >) - \sin(Jt < h_i^z >) \end{pmatrix} . \tag{5.10}$$

This is the generalization to quantum probability *amplitudes* of the probabilistic formulation of classical stochastic neurons. Indeed, the probabilities for the qubit to be in its eigenstates $\pm 1$ after a time $t$, obtained by squaring the probability amplitudes, are given by $f(\pm < h^z >)$, where $f(< h^z >) = (1 + \sin(2Jt < h^z >))/2$ has exactly the properties of an activation function (alternative to the Fermi function), at least in the region $Jt < \pi/4$. In this correspondence, the effective coupling constant $Jt$ plays the role of the inverse temperature, as usual in quantum mechanics. Correspondingly, thermal noise is substituted by quantum noise.

We shall now focus on a network without external inputs. In this case the equation for the average pattern overlaps has only the solution $|m| = 0$ for $0 < Jt < 1/2$. For such small effective couplings (high effective temperatures), corresponding to weak synaptic connections or to short evolution times, the network is unable to remember the stored pattern. For $1/2 < Jt$, however, the solution $|m| = 0$ becomes unstable, and two new stable solutions $m^z = \pm m_0$ appear: these are shown in Fig. 1 as a function of the effective coupling $Jt$. This means that the reaction of the mean orientation of the qubit spinors against a small deviation $\delta m^z$ from the $|m| = 0$ solution is larger than the deviation itself. Indeed, any so small external perturbation $(g/J)M^z$ present at the bifurcation time $t = 1/2J$ is sufficient for the network evolution to choose one of the two stable solutions, according to the sign of the external perturbation. The point $Jt = 1/2$ represents thus a quantum phase transition [11] from an amnesia (paramagnetic) phase to an ordered (ferromagnetic) phase in which the network has recall capabilities: the average pattern overlap $m^z$ is the corresponding order parameter. In the ferromagnetic phase the original $Z_2$ symmetry of the model is spontaneously broken (we will use this turn of words even if, strictly speaking, it refers to the choice of one out of many degenerate ground states rather than one of two possible quantum evolution paths related by a symmetry operation, the idea being exactly the same).

For $Jt = \pi/4$, the solution becomes $|m_0| = 1$, which means that the network is capable of perfect recall of the stored memory. For $Jt > \pi/4$ the

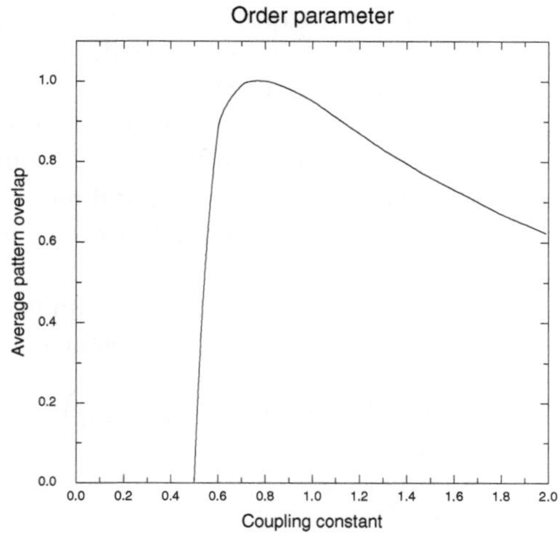


**Fig. 5.1.** The order parameter characterizing the quantum phase transition of networks with vanishing density of stored patterns.

solution $m_0$ decreases slowly to 0 again. Due to the periodicity of the time evolution, however, new stable solutions $m_0 = \pm 1$ appear at $Jt = (1+4n)\pi/4$ for every integer $n$. Also, for $Jt \geq 3\pi/4$, new solutions with $m^y \neq 0$ and $m^z = 0$ appear. These, however, correspond all to metastable states. Thus, $t = \pi/4J$ is the ideal computation time for the network.

The following picture of quantum associative memories emerges from the above construction. States of the network are generic linear superpositions of computational basis states. The network is prepared in the state $|\psi_0>$ and is then let to unitarily evolve for a time $t$. After this time the state of the network is measured, giving the result of the computation. During the evolution each qubit updates its quantum state by a rotation that depends on the aggregated synaptic potential determined by the state of all other qubits. These synaptic potentials are subject to large quantum fluctutations which are symmetric around the mean value $< h^z >= 0$. If the interaction is strong enough, any external disturbance will cause the fluctuations to collapse onto a collective rotation of all the network's qubits towards the nearest memory. Contrary to the spirit of the quantum circuit model, where maximally two-qubit interactions are switched on and off sequentially, in this model the qubits have long-range interactions which are never switched off.

## 5.4 Phase Structure

We will now turn to the more interesting case of a finite density $\alpha = p/n$ of stored memories in the limit $n \to \infty$. In this case the state of the network can have a finite overlap with several stored memories $\xi^\mu$ simultaneously. As in the classical case we shall focus on the most interesting case of a single "condensed pattern", in which the network uniquely recalls one memory without admixtures. Without loss of generality we will chose this memory to be the first, $\mu = 1$, omitting from now on the memory superscript on the corresponding overlap $m$. Correspondingly we will consider external inputs so that only $M^{\mu=1} = M \neq 0$. For simplicity of presentation, we will focus directly on solutions with a non-vanishing pattern overlap along the z-axis, omitting also the direction superscript $z$.

In case of a finite density of stored patterns, one cannot neglect the noise effect due to the infinite number of memories. This changes (5.8) to

$$\begin{aligned} m &= \frac{1}{n} \sum_i \sin 2Jt \left( m + \frac{g}{J} M + \Delta_i \right) , \\ \Delta_i &= \sum_{\mu \neq 1} \xi_i^1 \xi_i^\mu m^\mu . \end{aligned} \tag{5.11}$$

As in the classical case we will assume that $\{\xi_i^\mu\}$ and $\{m^\mu, \mu \neq 1\}$ are all independent random variables with mean zero and we will denote by square brackets the configurational average over the distributions of these random variables. As a consequence of this assumption, the mean and variance of the noise term are given by $[\Delta_i] = 0$ and $[\Delta_i^2] = \alpha r$, where

$$r = \frac{1}{\alpha} \sum_{\mu \neq 1} \left[ (m^\mu)^2 \right] \tag{5.12}$$

is the spin-glass order parameter [9]. According to the central limit theorem one can now replace $n^{-1} \sum_i$ in (5.11) by an average over a Gaussian noise,

$$m = \int \frac{dz}{\sqrt{2\pi}} \, \mathrm{e}^{\frac{-z^2}{2}} \sin 2Jt \left( m + \frac{g}{J} M + \sqrt{\alpha r} z \right) . \tag{5.13}$$

The second order parameter $r$ has to be evaluated self-consistently by a similar procedure starting from the equation analogous to eq. (5.11) for $\mu \neq 1$. In this case one can use $m^\mu \ll 1$ for $\mu \neq 1$ to expand the transcendental function on the right-hand side in powers of this small parameter, which gives

$$\begin{aligned} v &= \int \frac{dz}{\sqrt{2\pi}} \, \mathrm{e}^{\frac{-z^2}{2}} \sin^2 2Jt \left( m + \frac{g}{J} M + \sqrt{\alpha r} z \right) , \\ x &= \int \frac{dz}{\sqrt{2\pi}} \, \mathrm{e}^{\frac{-z^2}{2}} \cos 2Jt \left( m + \frac{g}{J} M + \sqrt{\alpha r} z \right) , \end{aligned} \tag{5.14}$$

where $v = (1 - 2Jtx)^2 r$. Solving the integrals gives finally the following coupled equations for the two order parameters $m$ and $r$:

$$\begin{aligned} m &= \sin 2Jt \left( m + \frac{g}{J} M \right) \, \mathrm{e}^{-2(Jt)^2 \alpha r} , \\ r &= \frac{1}{2} \, \frac{1 - \cos 4Jt \left( m + \frac{g}{J} M \right) \, \mathrm{e}^{-8(Jt)^2 \alpha r}}{\left( 1 - 2Jt \cos 2Jt \left( m + \frac{g}{J} M \right) \, \mathrm{e}^{-2(Jt)^2 \alpha r} \right)^2} . \end{aligned} \tag{5.15}$$

In terms of these order parameters one can distinguish three phases of the network. First of all the value of $m$ determines the presence ($m > 0$) or absence ($m = 0$) of ferromagnetic order (F). If $m = 0$ the network can be in a paramagnetic phase (P) if also $r = 0$ or a quantum spin glass phase (SG) if $r > 0$. The phase structure resulting from a numerical solution of the coupled equations (5.15) for $g = 0$ is shown in Fig. 2.

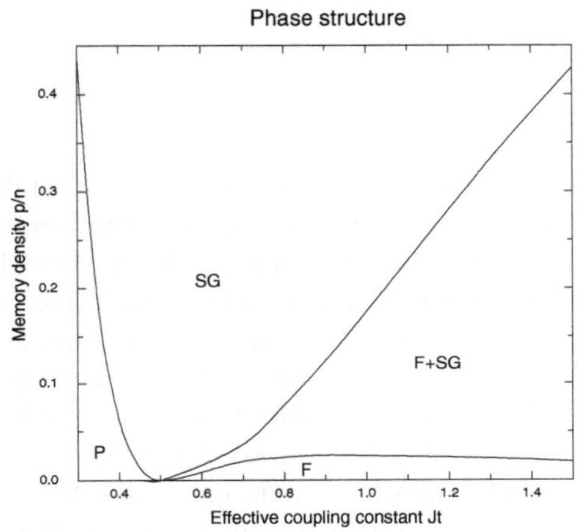


**Fig. 5.2.** The phase structure of the qubit network with finite density of stored memories. P, F and SG denote (quantum) paramagnetic, ferromagnetic and spin-glass phases, respectively. F + SG denotes a mixed phase in which the memory retrieval solution is only locally stable.

For $\alpha < 0.025$ the picture is not very different from the single memory case. For large enough computation times there exists a ferromagnetic phase in which the $m = 0$ solution is unstable and the network has recall capabilities. The only difference is that the maximum value of the order parameter $m$ is smaller than 1 (recall is not perfect due to noise) and the ideal computation time $t$ at which the maximum is reached depends on $\alpha$. For $0.025 < \alpha < 1.000$ instead, ferromagnetic order coexists as a metastable state with a quantum spin glass state. This means that ending up in the memory retrieval solution depends not only on the presence of an external stimulus but also on its magnitude; in other words, the external pattern has to be close enough to the stored memory in order to be retrieved. For $1 < \alpha$ all retrieval capabilities are lost and the network will be in a quantum spin glass state for all computation times (after the transition from the quantum paramagnet). $\alpha = 1$ is thus the maximum memory capacity of this quantum network. Note that $\alpha = 1$ corresponds to the maximum possible number of linearly independent memories. For memory densities smaller but close to this maximum value, however, the ferromagnetic solution exists only for a small range of effective couplings centered around $Jt \simeq 9$: for these high values of $Jt$ the quality of pattern retrieval is poor, the value of the order parameter $m$ being of the order 0.15-0.2. Much better retrieval qualities are obtained for smaller effective couplings: e.g. for $Jt = 1$ the order parameter is larger than 0.9 (corresponding to an error rate smaller than 5%) for memory densities up to 0.1. In this case, however the maximum memory density is 0.175, comparable with the classical result of the Hopfield model.

## 5.5 Summary

In summary, we have described a new model of quantum information processing that is capable of content association by quantum self-organization. This model can be viewed as the quantization of the classical Hopfield model of associative memory. The basic idea is to encode information in specific long-range qubit interactions and to let the corresponding quantum evolution amplify the desired result corresponding to a given input, a principle that lends itself to many generalizations. Straightforwardly, one could repeat the above analysis for other weights, e.g. the so-called pseudo-inverse solution [1] for linearly independent patterns or else one could ask the inverse question of finding the optimal weights to retrieve a given set of patterns. More generally one can also use the principles outlined here to construct and analyze quantum Boltzmann machines. Finally, it would be interesting to quantize not only the retrieval dynamics but also the memorized patterns, i.e. ask the question if it is possible to retrieve also linear superpositions of entangled patterns. In this case the present model could also be viewed as a quantum cloner (for N given, linearly independent patterns). Whatever the direction of future developments, however, the quantum associative memory model presented in this chapter is a first concrete example of the extension potential of the quantum information processing paradigm to "intelligent" tasks.

## References

1. For a review see: B. Müller and J. Reinhardt, Neural Networks, Springer-Verlag, Berlin, Germany 1990.
2. For a review see: M. A. Nielsen and I. L. Chuang, Quantum Computation and Quantum Information, Cambridge University Press, Cambridge, UK, 2000.
3. J.J. Hopfield, Proc. Natl. Acad. Scie. USA 79, p. 2554, 1982.
4. M. C. Diamantini and C. A. Trugenberger, Phys. Rev. Lett. 97, p. 130503, 2006.
5. M. Christandl, N. Datta, T. C. Dorlas, A. Ekert, A. Kay and A. J. Landahl, Phys. Rev. A71, p. 032312, 2005.
6. O. Mandel, M. Greiner, A. Widera, T. Rom, T. W. Hänsch and I. Bloch, Nature 425, p.937, 2003.
7. B. E. Kane, Nature 393, p.133, 2003.

8. W. S. McCullogh and W. Pitts, Bull. Math. Biophys. 5, p.115, 1943.
9. For a review see e.g. M. Mezard, G. Parisi and M. A. Virasoro, Spin Glass Theory and Beyond, World Scientific, Singapore, 1987.
10. See e.g. G. Parisi, Statistical Field Theory, Addison-Wesley, Redwood City, USA, 1988.
11. For a review see: S. Sachdev, Quantum Phase Transition, Cambridge University Press, Cambridge, UK 1999.

6

# Quantum-Inspired Evolutionary Algorithm for Numerical Optimization

André Vargas Abs da Cruz[1], Marley M. B. R. Vellasco[1], and Marco Aurélio C. Pacheco[1]

Pontifícia Universidade Católica do Rio de Janeiro
Rua Marquês de São Vicente, 225 - Gávea
CEP 22453-900 - Rio de Janeiro - Brasil
{andrev,marley,marco}@ele.puc-rio.br

Since they were proposed as an optimization method, evolutionary algorithms (EA) have been used to solve problems in several research fields. This success is due, besides other things, to the fact that these algorithms do not require previous information regarding the problem to be optimized and offer a high degree of parallelism. However, some problems are computationally intensive regarding the evaluation of each solution, which makes the optimization by EA's slow in some situations. This chapter proposes a novel EA for numerical optimization inspired by the multiple universes principle of quantum computing that presents faster convergence time for the benchmark problems. Results show that this algorithm can find better solutions, with less evaluations, when compared with similar algorithms, which greatly reduces the convergence time.

## 6.1 Introduction

Numerical optimization problems are an important field of research and have a wide number of applications, from the mathematical optimization of functions to the optimization of synaptic weights of a neural network.

Evolutionary algorithms have been an important tool for solving this sort of problems [2, 6]. The characteristics of dispensing a rigorous mathematical formulation regarding the problem one wants to solve, its high parallelism and easy of adaptation to several kinds of applications, make this class of algorithms very efficient to solve these problems.

Although they have been successful in solving several optimization problems, evolutionary algorithms present, in some specific problems, a slower performance than one could expect. This is due to slow evaluation functions (for instance, a complex simulator for an industrial plant) that can slow down the performance of the algorithm to unacceptable levels.

A.V. Abs da Cruz et al.: *Quantum-Inspired Evolutionary Algorithm for Numerical Optimization*, Studies in Computational Intelligence (SCI) **121**, 115–132 (2008)
www.springerlink.com 

To reduce this problem the quantum-inspired evolutionary algorithms [5,7] have been proposed as a novel approach in the field of evolutionary computation. These algorithms have been applied to combinatorial optimization problems [5] using a binary-based representation and have presented a better performance in solving this kind of problems than conventional algorithms.

For numerical optimization problems, however, a direct representation, where real numbers are directly encoded in a chromosome, is usually preferred rather than converting binary strings to numbers [6]. With a real number representation, the storage need is reduced while the numerical precision is increased.

This chapter presents a novel representation for the quantum-inspired genetic algorithms. This representation, based on real numbers, is a powerful tool for numerical optimization problems and is more efficient than conventional genetic algorithms that uses a real vector representation to optimize mathematical functions. Besides many other important properties, this model has the ability to find a good solution in a faster way using less individuals. This feature dramatically reduces the number of evaluations needed, which is an important performance factor when the model is being used in problems where each evaluation takes too much time to be completed.

This novel representation is based on the previous work about quantum-inspired algorithms for numerical optimization [3,4]. In this work, a simplified representation is used and the performance of the algorithm is tested with a new set of applications.

This work is divided in three additional sections. Section 2 details the proposed model. Section 3 discusses the results obtained in the optimization of a set of benchmark functions and in the optimization of the weights of a neural network used in a forecasting problem. Finally, section 4 presents the conclusions of this work.

## 6.2 The Quantum-Inspired Evolutionary Algorithm using a Real Number Representation

The algorithm 6.1 shows the complete quantum-inspired evolutionary algorithm using real number representation (QIEA). This proposed algorithm is explained in details in the following sections.

### 6.2.1 The Quantum Population

The quantum population $Q(t)$ is the QIEA's kernel. This population represents a superposition of states that is observed to generate classical individuals that are then, evaluated. The quantum population $Q(t)$ is made up of a set of $N$ quantum individuals $q_i$ $(i = 1, 2, 3, .., N)$. Each quantum individual $q_i$ is formed by $G$ genes $g_{ij}$ $(j = 1, 2, 3, ..., G)$ which consist of a pair of values, as shown in equation 6.1.

**Algorithm 6.1** Full quantum-inspired evolutionary algorithm using real representation pseudo-code.

1. $t \leftarrow 1$
2. Create quantum pop. $Q(t)$ with $N$ individuals with $G$ genes
3. **while** ($t <= T$)
4. Create the CDF's (Cumulative Distribution Function) using the quantum individuals
5. $E(t) \leftarrow$ generate classical pop. observing quantum pop. and using CDF's
6. **if** (t=1)
7. **then** $C(t) \leftarrow E(t)$
8. **else**
9. $E(t) \leftarrow$ *Crossover* between $E(t)$ and $C(t)$
10. evaluate $E(t)$
11. $C(t) \leftarrow K$ best individuals from $[E(t) + C(t)]$
12. **end if**
13. **with** the $N$ better individuals from $C(t)$
14. $Q(t+1) \leftarrow$ apply translate operation to $Q(t)$
15. $Q(t+1) \leftarrow$ apply resize operation to $Q(t+1)$
16. $t \leftarrow t+1$
17. **end with**
18. **end while**

$$q_i = [g_{ij}, g_{i2}, \ldots, g_{iG} | g_{ij} = (\rho_{ij}, \sigma_{ij})] \tag{6.1}$$

The values $\rho_{ij}$ and $\sigma_{ij}$ (where $\rho_{ij}, \sigma_{ij} \in \Re$) represent, respectively, the mean and the width of a square pulse, which is used by the algorithm to constrain the set of possible observable values inside the domain where the optimization should take place. In other words, each gene in the quantum individual represents an interval in the search space. The pulses' height $h_{ij}$ is calculated using the pulse's width value and the total number $N$ of quantum individuals in the population, as shown in equation 6.2.

$$h_{ij} = \frac{1/\sigma_{ij}}{N} \tag{6.2}$$

This equation guarantees that, later in the algorithm, the probability density functions used to generate classical individuals will have a total area equal to 1.

An example of a quantum gene $g_{ij} = (0, 2)$ is shown in figure 6.1 (in this example, we consider a quantum population formed by 2 individuals and thus, the height $h_{ij}$ of this gene is 0.25).

The creation of the quantum population, shown in step 2 in algorithm 6.1 is carried out by generating $N$ quantum individuals with random mean $\rho_{ij}$ inside the problem domain and with width $\sigma_{ij}$ equal to the total domain width. For example, considering that one wants to minimize the function $f(x)$ with

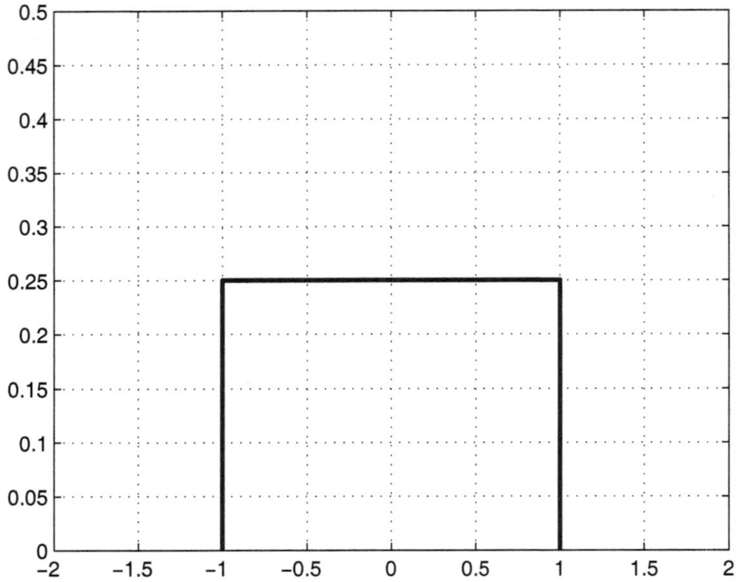


**Fig. 6.1.** An example of a quantum gene.

$x \in [-100, 100]$, the quantum individuals $q_1 = (0, 200)$, $q_2 = (-99.8, 200)$, $q_3 = (14.172637823, 200)$ and $q_4 = (100, 200)$ are examples of valid individuals in the initial population. It is important to notice that, usually, the square pulses in the initial population will be partially out of the domain bounds. In the latter example, the individual $q_1$ is the only one that is completely inside the initial domain bounds ($[-100, 100]$). The individual $q_2$, for instance, defines a square pulse in the interval $[-199.8, 0.2]$ being partially outside the domain bounds. However, this issue is not really a problem, as will become clear in the next steps of the algorithm.

### 6.2.2 Quantum Individuals Observation

After initializing the quantum population in step 2, the algorithm enters in the main evolutionary process loop. This loop will be executed for a total number $T$ of generations and is made up of several tasks.

Step 4 is one of the most importants in the algorithm. In this step, the interference process between the quantum individuals is executed to generate a Probability Density Function (PDF). This process basically consists in summing up the individuals that form the quantum population. In other words, in this step, the first gene of all individuals are summed. Next, the square pulses from the second gene for all individuals are also summed, and so on, until all genes from all individuals have been summed up. Mathematically, the probability density function of gene $j$, on generation $t$ of the algorithm, is given by the equation 6.3.

$$PDF_j = \sum_{i}^{N} g_{ij} \tag{6.3}$$

As an example, let's consider the quantum population to be formed by two individuals, each one made of 2 genes. The configuration of these individuals is given in table 6.1.

| Individual | Genes |
|---|---|
| $q_1$ | $g_{11} = (-5, 20)$, $g_{12} = (0, 20)$ |
| $q_2$ | $g_{21} = (5, 20)$, $g_{22} = (5, 20)$ |

**Table 6.1.** Example of quantum individuals forming population Q(t).

The graphical representation of these individuals is shown in figure 6.2. The resultant PDF's of the individuals shown in table 6.1 are shown in figure 6.3.

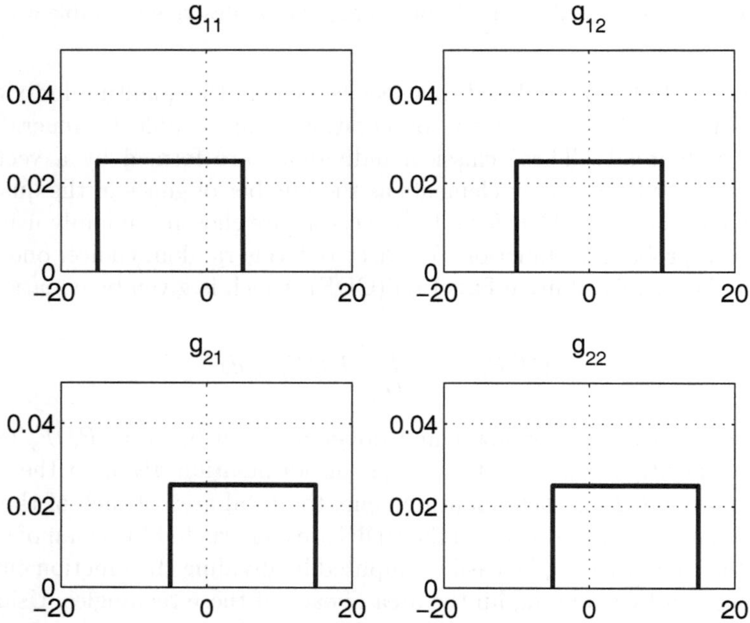


**Fig. 6.2.** Genes of 2 quantum individuals.

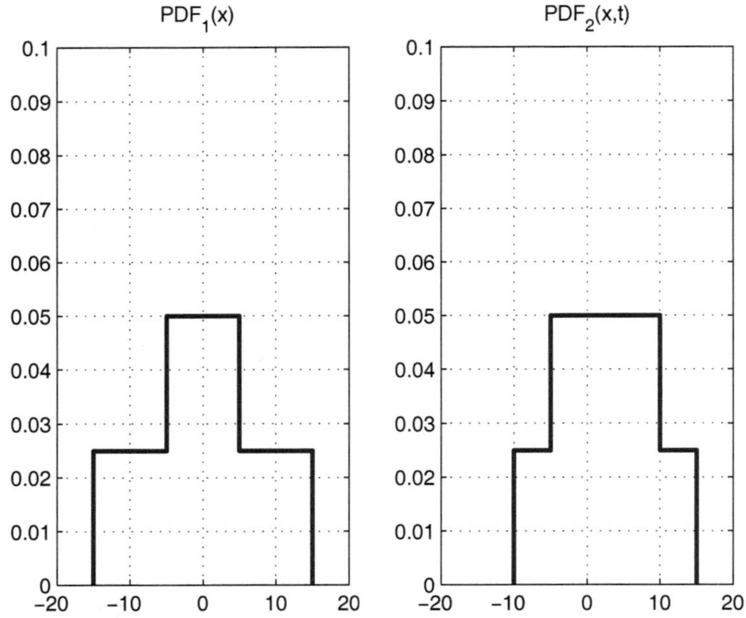


**Fig. 6.3.** The resultant PDFs of the population described in table 6.1.

Thus, the PDF's are related to a specific gene of the quantum individuals. It is from these PDF's that the observations will be able to generate the classical individuals. These classical individuals are formed by a vector of real numbers, with as many elements as the number of genes in the quantum individuals. The values that form these vectors are chosen randomly using the PDF's as a probability function. To carry out this random choice, one needs the Cumulative Distributive Function (CDF), which is given by equation 6.4.

$$CDF_j(x) = \int_{L_j^i}^{L_j^s} PDF_j(x)dx \tag{6.4}$$

Where $L_j^i$ e $L_j^s$ are the lower and upper limits of function $PDF_j$ respectively. As mentioned before, thanks to the formulation given to the height $h_{ij}$ in equation 6.2 for each quantum gene, the total area of each of the PDF curves $PDF_j$ is 1. In addition, as the PDFs are constructed by a sum of square pulses, the PDF area can be easily computed by dividing the function curve in rectangles and by summing up the area of each of these rectangles. Using the example from figure 6.3, each PDF can be divided in 3 rectangles, as shown in figure 6.4.

The CDFs can then be computed from these PDFs based on the rectangles R1, R2, R3, R4, R5 and R6. The resultant CDF of the first gene varies from 0 to 0.25 (total area of rectangle R1) in the interval $[-15, 5]$, from 0.25 to 0.75

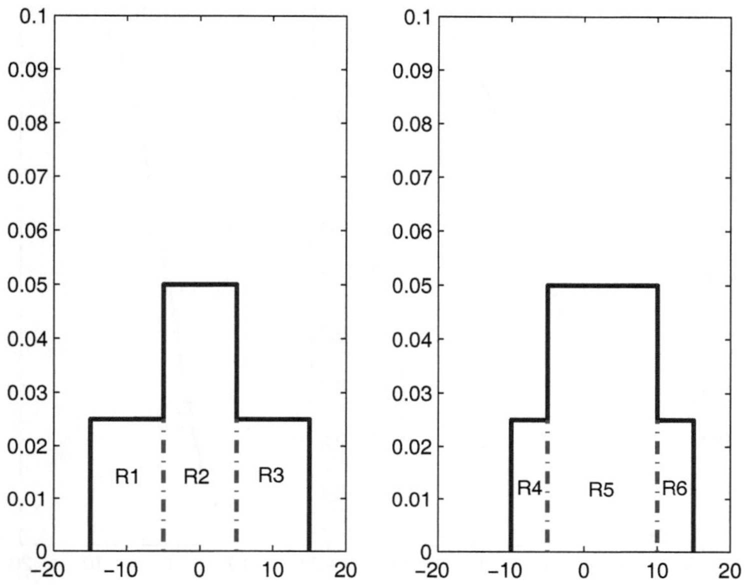


**Fig. 6.4.** PDF's division in rectangles.

(total area of rectangle R1 + rectangle R2) and from 0.75 to 1.0 (total curve area). The CDFs can be represented graphically as shown in figure 6.5.

After generating the CDFs, it is possible to generate a set of classical individuals by using these curves (step 5 from the algorithm). This classical population is created by choosing a random number between 0 and 1 and by identifying this point in the CDF. Mathematically, the value of a gene in the classical population is generated as shown in equation 6.5.

$$x = CDF^{-1}(r) \tag{6.5}$$

Where $r$ is a random number in the $[0, 1]$ interval. Through this process, a temporary classical population $E(t)$, with $K$ individuals is, then, automatically generated. As explained before, the domain of the PDFs (and so of the CDFs) can be outside the domain bounds of the problem. If the chosen value of a gene is outside the domain bounds, it is corrected by making it equal to the nearest domain bound.

If the algorithm is in the first generation (step 6), the classical population $C(t)$ is an exact copy of the temporary classical population $E(t)$ created in step 5. If the algorithm is not in the first generation, a crossover operator must be applied among the individuals of the temporary classical population $E(t)$ and the classical population $C(t)$. The crossover operator proposed in this work is explained in the algorithm 6.2.

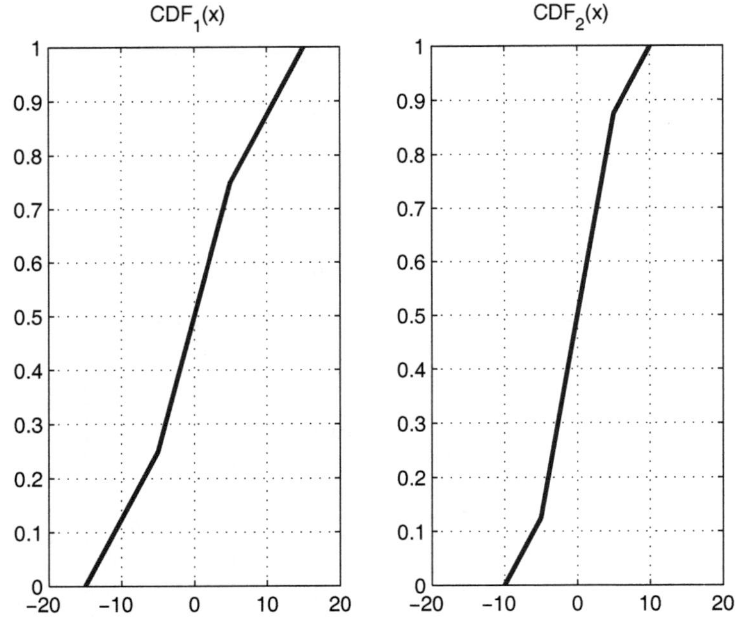


**Fig. 6.5.** CDFs examples.

---

**Algorithm 6.2** Crossover algorithm.

---

**for** i = 1 to K<br>
    select individual $e_i$ from $E(t)$<br>
    select individual $c_i$ from $C(t)$<br>
    **for** j = 1 to G<br>
        $r \leftarrow$ choose random number in $[0, 1)$ using an uniform distribution<br>
        **if** $r < \xi$<br>
            $e'_{ij} \leftarrow e_{ij}$<br>
        **else**<br>
            $e'_{ij} \leftarrow c_{ij}$<br>
        **end if**<br>
    **end for**<br>
**end for**

---

In this algorithm, $K$ is the number of individuals in classical population $C(t)$, $G$ is the number of genes in each individual and $\xi$ is the crossover rate used during the process. A rate equal to 1 carries on all the genes in the created individual to the offspring. A rate equal to 0 does not modify the individuals and no evolution is observed. An example of how this algorithm works can be seen in figure 6.6, where the sampled random numbers $r$ are $(0.9, 0.8, 0.1, 0.2, 0.7, 1.0, 0.15, 0.1)$ (the first number in the sequence is the random number for the first element in the vectors, the second number is the random number for the second element and so on) and the crossover rate is $\xi = 0.6$.

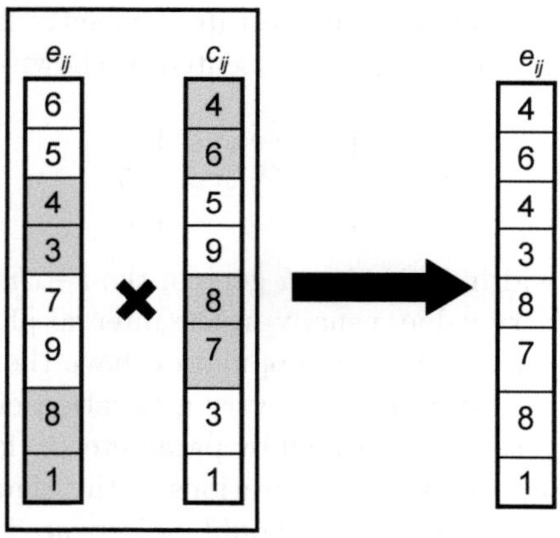


**Fig. 6.6.** Crossover Example.

It is worth pointing out that for the crossover to work properly, the number of temporary classical individuals created from the CDFs must be equal to the number $K$ of individuals in population $C(t)$.

After using the crossover operator, the temporary population $E(t)$ must be evaluated. Then, the individuals in this population and in population $C(t)$ are ordered in a single set and the $K$ best individuals from this new set are selected to form the new population $C(t)$.

### 6.2.3 Updating the Quantum Population

After generating the classical individuals, it is necessary to update the quantum individuals in the population $Q(t)$. This procedure is done in two steps. The first step (step 14 in the algorithm) modifies the center $\rho$ of the quantum genes. This procedure is simple and is done by making the mean values of each gene equal to the values of the genes from the classical individuals. Mathematically, this is represented by equation 6.6.

$$\rho_{ij} = c_{ij} \tag{6.6}$$

Where $\rho_{ij}$ is the value that defines the center of the $j-th$ gene of the $i-th$ quantum individuals in $Q(t)$ and $c_{ij}$ is the value of the $j-th$ gene of the $i-th$ classical individual in $C(t)$.

The second step in the process of updating the quantum population (step 15 of the algorithm) consists in increasing or decreasing the width of the quantum genes. This change in the width is done in an homogeneous way for all the quantum genes and for all the quantum individuals. The heuristic used to determine if the new width should be increased or decreased is the $1/5th$ rule [6]: if at most 20% of the classical population created in the current generation has improved, the gene width is decreased; if this rate is higher than 20% the width is increased; if the rate is exactly 20%, no changes are made. Mathematically, this can be represented as shown in equation 6.7.

$$\sigma_{ij} = \begin{cases} \sigma_{ij} \cdot \delta & \varphi < 1/5 \\ \sigma_{ij}/\delta & \varphi > 1/5 \\ \sigma_{ij} & \varphi = 1/5 \end{cases} \tag{6.7}$$

Where $\sigma_{ij}$ is the width of the $j-th$ gene of the $i-th$ quantum individual in $Q(t)$, $\delta$ is an arbitrary value, usually in the interval $[0,1]$ and $\varphi$ is the rate of how many individuals of the new population have their overall evaluation improved. This rule can be applied at every generation of the algorithm or it can be applied at larger steps, specified by parameter $k$ (for instance, if $k=5$, this rule will be applied every five generations of the algorithm).

All these steps of the algorithm should be repeated for a number $T$ of generations, according to what was shown in algorithm 6.1.

## 6.3 Case Studies

In this section, two different case studies are presented: in the first one, we apply the QIEA to the optimization of 4 benchmark functions; in the second one, we apply the QIEA to a neuro-evolution problem, where we want to optimize the weights of a neural network in a forecasting problem.

### 6.3.1 Optimization of Benchmark Functions

The benchmark functions used here is a set of 4 different benchmark functions selected from [8]. These functions are widely used as benchmark in numerical optimization and the cited paper allows the comparison of the proposed method with the following methods:

- Stochastic Genetic Algorithms (StGA) [8];
- Fast Evolutionary Programming (FEP) [10];
- Fast Evolutionary Strategy (FES) [9];

- Particle Swarm Optimization (PSO) [1].

All these functions should be minimized and are describe in table 6.2. Function $f_1$ is unimodal and relatively easy to minimize but, for higher dimensions, they become harder to optimize. Functions $f_2$, $f_3$ and $f_4$ are multimodal functions and have lots of local minima, representing a harder class of functions to optimize. The minimum of all these functions is 0. Figures 6.7, 6.8, 6.9 and 6.10 shows a graphical view of these functions with 2 variables.

| Equation | Domain |
|---|---|
| $f_1(\mathbf{x}) = \sum_{j=1}^{30} x_j^2$ | $x_j \in [-30, 30]$ |
| $f_2(\mathbf{x}) = \sum_{j=1}^{30} \lvert x_j \rvert + \prod_{j=1}^{30} \lvert x_j \rvert$ | $x_j \in [-10, 10]$ |
| $f_3(\mathbf{x}) = \frac{1}{4000} \sum_{j=1}^{30} {x_j}^2 - \prod_{j=1}^{30} \cos\left(\frac{x_j}{\sqrt{i}}\right) + 1$ | $x_j \in [-600, 600]$ |
| $f_4(\mathbf{x}) = -20 + \exp\left(-0.2\sqrt{\frac{1}{30}\sum_{j=1}^{30} x_j^2}\right) - \exp\left(\frac{1}{30}\sum_{j=1}^{30} \cos 2\pi x_j\right)$ | $x_j \in [-10, 10]$ |

**Table 6.2.** Functions to be optimized.

The parameter's configuration used to test the QIEA model for all the benchmark functions is given on table 6.3.

| | $f_1$ | $f_2$ | $f_3$ | $f_4$ |
|---|---|---|---|---|
| Quantum Population $Q(t)$ Size | 10 | 5 | 5 | 4 |
| Classical Observed Population Size | 10 | 5 | 10 | 4 |
| Number of generations $k$ before checking improvement | 10 | 8 | 5 | 20 |
| Crossover Rate | 0.1 | 0.1 | 0.1 | 1 |
| Number of Generations | 3000 | 3520 | 5250 | 2500 |

**Table 6.3.** Parameters for the QIEA.

All functions were optimized for 50 experiments. The number of generations was chosen in order to make the total number of evaluations equal to the number of evaluations for the Stochastic Genetic Algorithm in [8], which is the comparison method that has shown the best results. The mean number of evaluations for each experiment is shown on table 6.4.

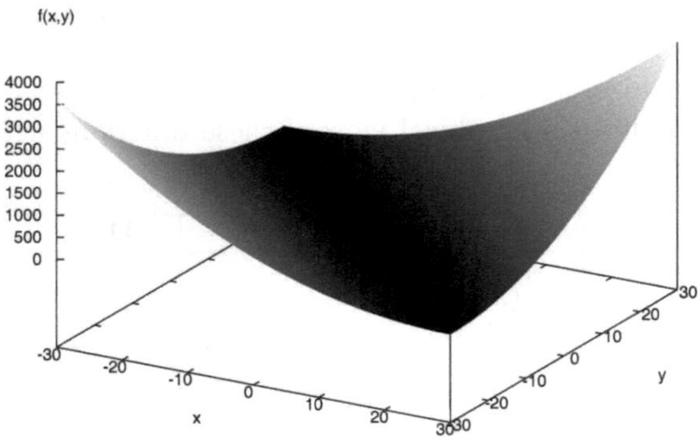


**Fig. 6.7.** Function $f_1$.

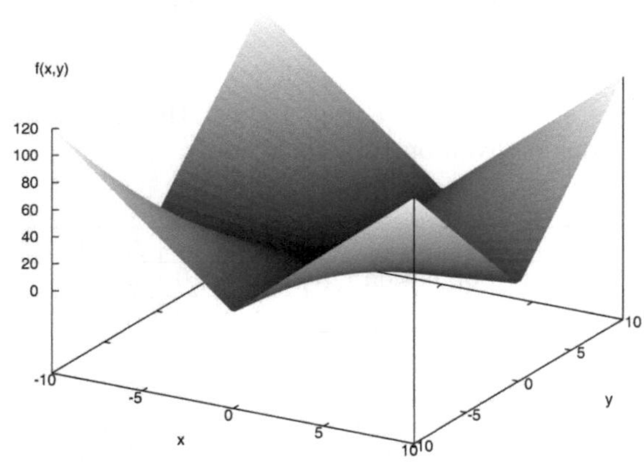


**Fig. 6.8.** Function $f_2$.

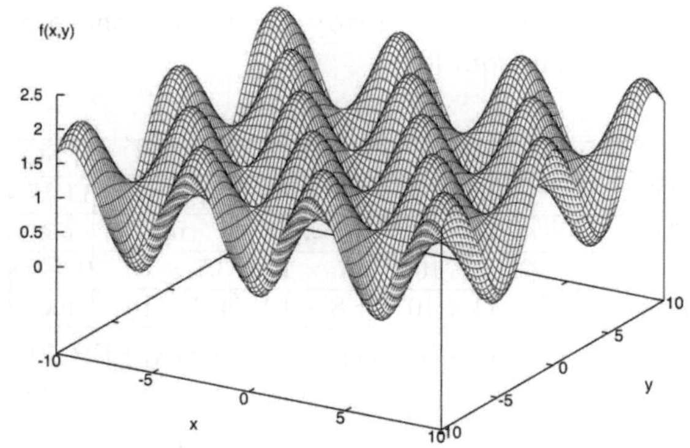


**Fig. 6.9.** Function $f_3$.

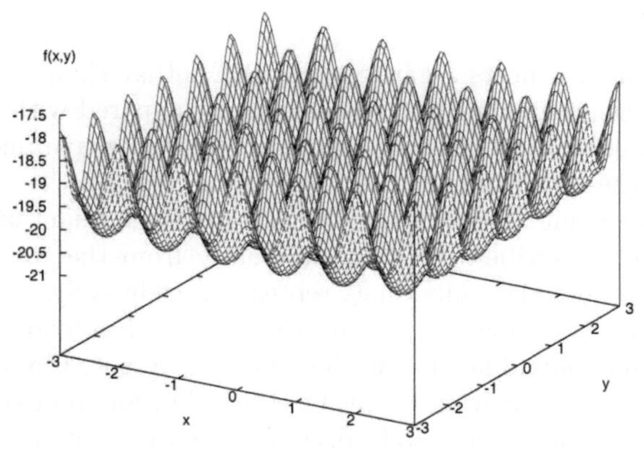


**Fig. 6.10.** Function $f_4$.

| | QIEA | StGA | FEP | FES | PSO |
|---|---|---|---|---|---|
| $f_1$ | 30000 | 30000 | 150000 | n.a. | 250000 |
| $f_2$ | 17600 | 17600 | 200000 | n.a. | n.a. |
| $f_3$ | 52500 | 52500 | 200000 | 200030 | 250000 |
| $f_4$ | 10000 | 10000 | 150000 | 150030 | n.a. |

**Table 6.4.** Mean Number of Function Evaluations for each experiment.

The results obtained with the QIEA model, as well as the results attained by the other models, are shown in table 6.5 with respect to the mean best value found for each function. As already mentioned, the results for all the other algorithms are taken from [8].

| | QIEA | StGA | FEP | FES | PSO |
|---|---|---|---|---|---|
| $f_1$ | $2.0 \times 10^{-15}$ | $2.45 \times 10^{-15}$ | $5.7 \times 10^{-4}$ | n.a. | 11.175 |
| $f_2$ | $7.6 \times 10^{-10}$ | $2.03 \times 10^{-7}$ | $8.1 \times 10^{-3}$ | n.a. | n.a. |
| $f_3$ | 0 | $2.44 \times 10^{-17}$ | $1.6 \times 10^{-2}$ | $3.7 \times 10^{-2}$ | 0.4498 |
| $f_4$ | $1.36 \times 10^{-8}$ | $3.52 \times 10^{-8}$ | $1.8 \times 10^{-2}$ | $1.2 \times 10^{-2}$ | n.a. |

**Table 6.5.** Comparison results between QIEA, StGA, FEP, FES and PSO.

As can be observed from table 6.5, the QIEA model obtained better results with much less evaluations than FEP, FES and PSO. Additionally, the algorithm was able to find better results than StGA with the same number of function evaluations.

### 6.3.2 Discussion

There are two main aspects regarding the QIEA that should be discussed: the faster convergence of the QIEA model when compared with conventional genetic algorithms and the impact of the configuration parameters on the performance of the algorithm.

The first important aspect of the QIEA is that the quantum population is able to directly describe schemata, differently from the conventional genetic algorithm where the individuals represent, exclusively, a set of points in the search space. Schemata represent a whole set of individuals; thus, the evaluation of some individuals inside the region represented by the schemata provides, not only an evaluation for each of them, but an approximate evaluation of the whole region of the search space that is covered by the schemata (as long as this region does not have abrupt changes). The ability of evaluating whole regions of the search space instead of single points allows the algorithm to identify the promising regions of the search space faster than conventional genetic algorithms.

Another way of understanding how the QIEA is capable of converging is through an analogy with evolutionary strategies [6]. In this kind of algorithm, the population is formed by a single individual that consists of a vector of real numbers $\mathbf{v}$ and a vector of variances $\boldsymbol{\sigma}$. The only operator in evolutionary strategies is a mutation operator that changes the genes in $\mathbf{v}$ by a small quantity $\delta_i$, chosen randomly from a Gaussian distribution with mean zero and variance $\sigma_i$ (where $i$ is the index of each one of the genes that make $\mathbf{v}$). The quantum population can be seen as the individual from the evolutionary strategy, however, in this case, the set of individuals that forms the quantum population is capable of forming more complex curves than a Gaussian shape. Besides, since the quantum population is formed by several independent pulses, the curves can split among several regions of the search space. As the evolution goes on, the more promising regions are identified (as the tendency is that these regions will create more individuals with better evaluations) and the pulses are "attracted" to these regions, converging to an optimum solution. These characteristics help the algorithm to avoid being trapped in local minima and also speed up the algorithm.

The other important aspect to be discussed regards the parameters of the QIEA and how they influence the evolutionary process. There is no "magic" formula for choosing these values and experimentation is the most trustable way to identify them. However, some important observations can be made regarding the parameters.

In relation to the number of individuals in the quantum population, it is important to take to consideration that a very small number of individuals can lead to a premature convergence of the algorithm. In this case, the pulses that form the population will quickly move to a local minima and will be locked in these regions indefinitely. On the other hand, a large number of quantum individuals will increase the width of the sum of PDFs. Thus, the algorithm will have problems to converge to the optimal solution, even if it is able to find the region where the optimal point is. The differences between the smoothness of the PDFs' sum are shown in figure 6.11 and 6.12, where the first one is the result of a population of 5 quantum individuals and the latter of 20 individuals.

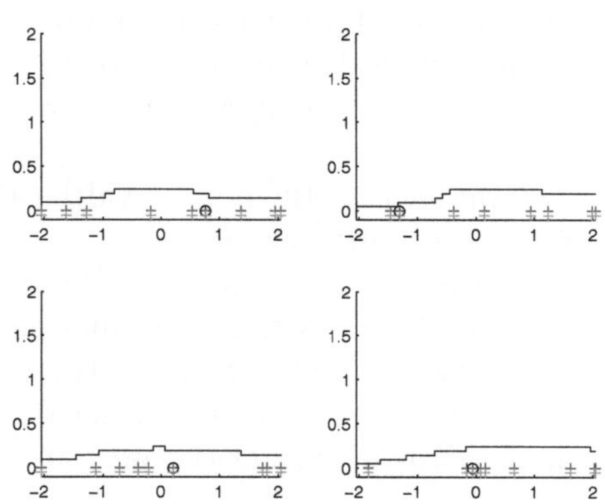

**Fig. 6.11.** Sum of PDFs of a population with 5 quantum individuals.

The size of the classical population (the observed population) is not critical for the algorithm performance. Few individuals in a generation are, usually, enough for the algorithm to converge. Too many individuals will not make a big difference in the performance and will increase the number of evaluations, reducing the overall performance. Thus, it is recommended to use a small

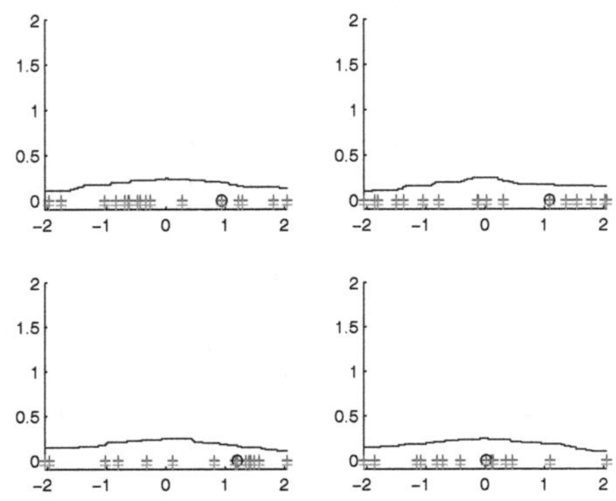

**Fig. 6.12.** Sum of PDFs of a population with 20 quantum individuals.

number of individuals. However, it is important to remember that this number should be, at least, equal to the number of quantum individuals.

The number of generations that must be created before the improvement rate of the algorithm is (parameter $k$ described in section 6.2.3) verified is also very important and should also be found by a process of experimentation. Each function that one wishes to optimize has a different behaviour regarding this parameter. Small values makes the pulses width vary very quickly, which, in some cases, make the time the algorithm has to explore a given region too small to allow it to find good individuals. On the other hand, large values make the algorithm lose to much time exploring a given region. Figure 6.13 shows how the pulses' width vary during the evolutionary process when the values of $k$ are 2, 5 and 10. As can be observed, the curves are smoother with higher values of $k$.

Figure 6.14 shows the variation of the pulses' width (with $k = 10$) and the improvement rate during the evolutionary process.

Finally, the crossover rate also has an important role in the evolutionary process. Usually, unimodal problems, with soft evaluation landscapes and few variables, can use a large crossover rate (values between 0.9 and 1.0 should be used for the first trials). Problems with lots of local minima or too much variables should use smaller crossover rates (between 0.1 and 0.2).

## 6.4 Conclusions and Future Works

This work presented a new quantum-inspired evolutionary algorithm with real number representation that is better suited for numerical optimization problems than using binary representation.

The new algorithm has been evaluated in several benchmark problems, in particular, for function optimization and supervised learning problems,

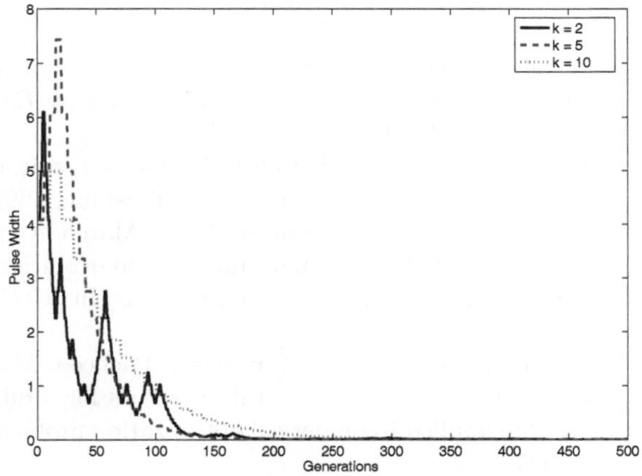


**Fig. 6.13.** How the pulses' width vary during the evolutionary process with different values of $k$.

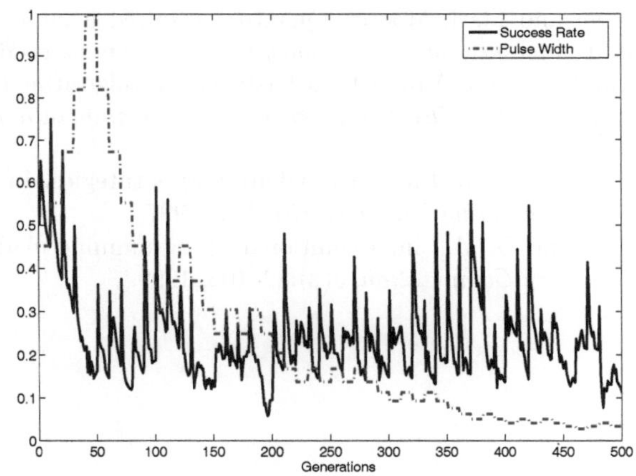


**Fig. 6.14.** Variation of pulses' width and the improvement rate during the evolutionary process.

and showed very promising results, with better performance than other well-established algorithms.

Future works include the use of the algorithm in neuroevolution control and other practical numerical optimization problems.

## References

1. P. J. Angeline. Evolutionary optimization versus particle swarm optimization: Philosophy and performance differences. In *Proceedings on Evolutionary Programming VII*, pages 601–610, 1998.
2. Thomas Back, David B. Fogel, and Zbigniew Michalewicz, editors. *Handbook of Evolutionary Computation.* Institute of Physics Publishing, 1997.
3. André V. Abs da Cruz, Carlos R. Hall Barbosa, Marco Aurélio Cavalcanti Pacheco, and Marley B. R. Vellasco. Quantum-inspired evolutionary algorithms and its application to numerical optimization problems. In *ICONIP*, pages 212–217, 2004.
4. André V. Abs da Cruz, Marco Aurélio Cavalcanti Pacheco, Marley B. R. Vellasco, and Carlos R. Hall Barbosa. Cultural operators for a quantum-inspired evolutionary algorithm applied to numerical optimization problems. In *IWINAC (2)*, pages 1–10, 2005.
5. K. Han and J. Kim. Quantum-inspired evolutionary algorithm for a class of combinatorial optimization. *IEEE Trans. Evol. Comput.*, 6(6):580–593, 2002.
6. Zbigniew Michalewicz. *Genetic algorithms + data structures = evolution programs (2nd, extended ed.).* Springer-Verlag New York, Inc., New York, NY, USA, 1994.
7. Ajit Narayanan and Mark Moore. Quantum inspired genetic algorithms. In *International Conference on Evolutionary Computation*, pages 61–66, 1996.
8. Zhenguo Tu and Yong Lu. A robust stochastic genetic algorithm (stga) for global numerical optimization. *IEEE Trans. Evolutionary Computation*, 8(5):456–470, 2004.
9. X. Yao, Y. Liu, and G. M. Lin. Fast evolutionary strategies. In *Proceedings on Evolutionary Programming VI*, pages 151–161, 1997.
10. X. Yao, Y. Liu, and G. M. Lin. Evolutionary programming made faster. *IEEE Trans. Evolutionary Computation*, 3(2):82–102, 1999.

# 7

# Calibration of the *VGSSD* Option Pricing Model using a Quantum-inspired Evolutionary Algorithm

Kai Fan[1,2], Conall O'Sullivan[2], Anthony Brabazon[1], Michael O'Neill[1], and Seán McGarraghy[2]

[1] Natural Computing Research and Applications Group, University College Dublin, Ireland.
kai.fan@ucd.ie; anthony.brabazon@ucd.ie; m.oneill@ucd.ie
[2] School of Business, University College Dublin, Ireland.
conall.osullivan@ucd.ie; sean.mcgarraghy@ucd.ie

Quantum effects are a natural phenomenon and just like evolution, the brain, or immune systems, can serve as an inspiration for the design of computing algorithms. This chapter illustrates how a quantum-inspired evolutionary algorithm (QIEA) using real number encodings can be constructed and examines the utility of the resulting algorithm on an important real-world problem, namely the calibration of an Option Pricing model. The results from the algorithm are shown to be robust and comparable to those of other algorithms, suggesting that there is useful potential to apply QIEA to this domain.

## 7.1 Introduction

This chapter introduces a novel option pricing calibration methodology which uses a quantum-inspired real number encoding rather than a traditional encoding representation in an evolutionary algorithm. The chapter also assesses the utility of the resulting algorithm for the purposes of calibrating an option pricing model.

Quantum mechanics is an extension of classical mechanics which models behaviours of natural systems that are observed particularly at very short time or distance scales. An example of such a system is a sub-atomic particle, such as a free electron. A complex-valued (deterministic) function of time and space co-ordinates, called the *wave-function*, is associated with the system: it describes the *quantum state* the system is in. The standard interpretation of Quantum Mechanics is that this abstract wave-function allows us to calculate

K. Fan et al.: *Calibration of the* VGSSD *Option Pricing Model using a Quantum-inspired Evolutionary Algorithm*, Studies in Computational Intelligence (SCI) **121**, 133–153 (2008)
www.springerlink.com

probabilities of outcomes of concrete experiments. The squared modulus of the wave-function is a probability density function (PDF): it describes the probability that an observation of, for example, a particle will find the particle at a given time in a given region of space. The wave-function satisfies the *Schrödinger equation.* This equation can be thought of as describing the time evolution of the wave-function — and so the PDF — at each point in space: as time goes on, the PDF becomes more "spread out" over space, and our knowledge of the position of the particle becomes less precise, until an observation is carried out; then, according to the usual interpretation, the wave-function "collapses" to a particular classical state (or *eigenstate*), in this case a particular position, and the spreading out of the PDF starts all over again.

Before the observation we may regard the system as being in a linear combination of all possible classical states (this is called *superposition of states*); then the act of observation causes one such classical state to be chosen, with probability given by the PDF. Note that the wave function may interfere with itself (for example, if a barrier with slits is placed in the "path" of a particle) and this interference may be constructive or destructive, that is, the probability of detecting a particle in a given position may go up or go down.

More generally, we may seek to observe properties of quantum systems other than position, e.g., energy, momentum, or the quantum *spin* of an electron, photon or other particle. [Spin is incorporated by necessity in Dirac's relativistic extension of the wave equation; in fact spin is one of the arguments of the wave-function.] Such properties are called *observables.* Observables may be either continuous (e.g., position of a particle) or discrete (e.g., the energy of an electron in a bound state in an atom). Some observables may only take finitely many values, e.g., there are only two possible values for a given particle's spin: "up" or "down". This last is an example of a *two-state system*: in such a system the quantum state $\psi$ is a linear superposition of just two eigenstates, say $|0\rangle$ and $|1\rangle$ in the standard Dirac bra-ket notation, that is,

$$\psi = \alpha|0\rangle + \beta|1\rangle,$$

where $\alpha$ and $\beta$ are complex numbers with $|\alpha|^2 + |\beta|^2 = 1$. Here $|0\rangle$ and $|1\rangle$ are basis vectors for a 2-dimensional complex Hilbert space. A two-state system where the states are normalised and orthogonal, as here, may be regarded as a *quantum bit* or *qubit.*[3] It is thought of as being in eigenstates $|0\rangle$ and $|1\rangle$ simultaneously, until an observation is made and the quantum state collapses to $|0\rangle$ (with probability $|\alpha|^2$) or $|1\rangle$ (with probability $|\beta|^2$). The relation $|\alpha|^2 + |\beta|^2 = 1$ captures the fact that precisely one of $|0\rangle$, $|1\rangle$ must be observed, so their probabilities of observation must sum to 1.

A *quantum computer* is one which works with qubits instead of the (classical) bits used by usual computers. Benioff [1] first considered a Turing machine

[3] Geometrically, a qubit is a compact 2-dimensional complex manifold, called the Bloch sphere.

which used a tape containing what we would call qubits. Feynman [10] developed examples of physical computing systems not equivalent to the standard model of deterministic computation, the Turing machine.

In recent years there has been a substantial interest in the theory and design of quantum computers, and the design of programs which could run on such computers, stimulated by Shor's discovery of a quantum factoring algorithm which would run faster than possible clasically. One interesting strand of research has been the use of natural computing (for example Genetic Programming (GP)) to generate quantum circuits or programs (algorithms) for quantum computers [25]. (Genetic programming is an evolutionary algorithm based methodology inspired by biological evolution to find computer programs that perform a user-defined task. Therefore it is a machine learning technique used to optimize a population of computer programs according to a fitness landscape determined by a program's ability to perform a given computational task.) There has also been associated work in a reverse direction which draws inspiration from concepts in quantum mechanics in order to design novel natural computing algorithms. This is currently an area of active research interest. For example, quantum-inspired concepts have been applied to the domains of evolutionary algorithms [12, 14, 23, 27, 28], social computing [29], neuro-computing [11,18,26], and immuno-computing [16,19]. A claimed benefit of these algorithms is that because they use a quantum representation, they can maintain a good balance between exploration and exploitation. It is also suggested that they offer computational efficiencies as use of a quantum representation can allow the use of smaller population sizes than typical evolutionary algorithms.

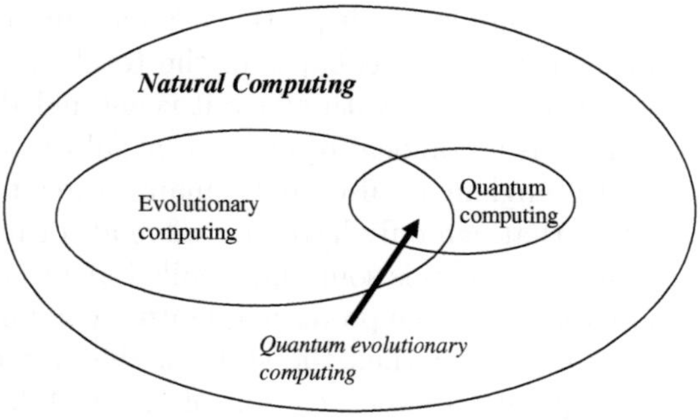


**Fig. 7.1.** Quantum-inspired evolutionary computing

Quantum-inspired evolutionary algorithms (QIEA) offer interesting potential. As yet, due to their novelty, only a small number of recent papers have implemented a QIEA, typically reporting good results [27,28]. Consequently, we have a limited understanding of the performance of these algorithms and further testing is required in order to determine both their effectiveness and their efficiency. It is also noted that although a wide-variety of biologically-inspired algorithms have been applied for financial modelling [3], the QIEA methodology has not yet been applied to the finance domain. This study addresses both of these research gaps.

### Structure of Chapter

The rest of this chapter is organised as follows. The next section provides a concise overview of QIEA, concentrating on the quantum-inspired genetic algorithm. We then outline the experimental methodology adopted. The remaining sections provide the results of these experiments followed by a number of conclusions.

## 7.2 The Quantum-inspired Genetic Algorithm

The best-known application of quantum-inspired concepts in evolutionary computing is the quantum-inspired genetic algorithm (QIGA) [12, 14, 23, 27, 28]. The QIGA is based on the concepts of a qubit and the superposition of states. In essence, in QIGAs the traditional representations used in evolutionary algorithms (binary, numeric and symbolic) are extended to include a quantum representation.

A crucial difference between a qubit and a (classical) bit is that multiple qubits can exhibit quantum entanglement. Entanglement is when the wave function of a system composed of many particles cannot be separated into independent wave functions, one for each particle. A measurement made on one particle can produce, through the collapse of the total wavefunction, an instantaneous effect on other particles with which it is entangled, even if they are far apart. Entanglement is a nonlocal property that allows a set of qubits to be highly correlated. Entanglement also allows many states to be acted on simultaneously, unlike bits that can only have one value at a time. The use of entanglement in quantum computers is sometimes called *quantum parallelism*, and gives a possible explanation for the power of quantum computing: because the state of the quantum computer (i.e., the state of the system considered as a whole) can be in a quantum superposition of many different classical computational states, these classical computations can all be carried out at the same time.

The quantum equivalent of a classical operator on bits is an *evolution* (not to be confused with the evolution of EAs). It transforms an input to an output, e.g., by rotation or Hadamard gate, and operates without measuring the value of the qubit(s). Thus it effectively does a parallel computation on all the qubits at once and gives rise to a new superposition.

In the language of evolutionary computation a system of $m$ qubits may be referred to as a *quantum chromosome* and can be written as a matrix with two rows:

$$\begin{bmatrix} \alpha_1 \ \alpha_2 \ \ldots \ \alpha_m \\ \beta_1 \ \beta_2 \ \ldots \ \beta_m \end{bmatrix}. \tag{7.1}$$

A key point when considering quantum systems is that they can compactly convey information on a large number of possible system states. In classical bit strings, a string of length $m$ can represent $2^m$ possible states. However, a quantum space of $m$ qubits has $2^m$ *dimensions* (as a complex manifold).[4] Thus, a single qubit register of length $m$ can simultaneously represent *all* possible bit strings of length $2^m$, e.g., an 8 qubit system can simultaneously encode 256 distinct strings. This implies that it is possible to modify standard evolutionary algorithms to work with very few, or even a single quantum individual, rather than having to use a large population of solution encodings. The qubit representation can also help to maintain diversity during the search process of an evolutionary algorithm, due to its capability to represent multiple system states simultaneously.

### 7.2.1 Representing a Quantum System

There are many ways that a quantum system could be defined in order to encode a set of binary (solution) strings. For example, in the following 3 qubit quantum system, the quantum chromosome is defined using the three pairs of amplitudes below

$$\begin{bmatrix} \frac{1}{\sqrt{2}} & \frac{\sqrt{3}}{2} & \frac{1}{2} \\ \frac{1}{\sqrt{2}} & \frac{1}{2} & \frac{\sqrt{3}}{2} \end{bmatrix} \tag{7.2}$$

These numbers are the probabilities that a qubit (unit of information) will be observed in a particular eigenstate rather than another. Taking the first qubit, the occurrence of either state 0 or 1 is equally likely as both $\alpha_1$ and $\beta_1$ have the same amplitude. Following on from the definition of the 3 qubit system, the (quantum) state of the system is given by

$$\tfrac{\sqrt{3}}{4\sqrt{2}}|000\rangle+\tfrac{3}{4\sqrt{2}}|001\rangle+\tfrac{1}{4\sqrt{2}}|010\rangle+\tfrac{\sqrt{3}}{4\sqrt{2}}|011\rangle+\tfrac{\sqrt{3}}{4\sqrt{2}}|100\rangle+\tfrac{3}{4\sqrt{2}}|101\rangle+\tfrac{1}{4\sqrt{2}}|110\rangle+\tfrac{\sqrt{3}}{4\sqrt{2}}|111\rangle \tag{7.3}$$

To provide intuition on this point, consider the system state $|000\rangle$. The associated probability amplitude for this state is $\frac{\sqrt{3}}{4\sqrt{2}}$ and this is derived from the probability amplitudes of the 0 state for each of the three individual qubits ($\frac{1}{\sqrt{2}} * \frac{\sqrt{3}}{2} * \frac{1}{2} = 0.25$). The associated probabilities of each of the individual states ($|000\rangle, |001\rangle, |010\rangle, |011\rangle, |100\rangle, |101\rangle, |110\rangle, |111\rangle$) are $\frac{3}{32}, \frac{9}{32}, \frac{1}{32}, \frac{3}{32}, \frac{3}{32}, \frac{9}{32}, \frac{1}{32}, \frac{3}{32}$ respectively. Taking the first of these states as an example, $(\frac{\sqrt{3}}{4\sqrt{2}})^2 = \frac{3}{32}$.

[4] It can be shown that, because of entanglement, an $m$-qubit physical system has $2^{m+1} - 2$ degrees of freedom, much larger than the $2m$ degrees a classical version would have.

### 7.2.2 Real-valued quantum-inspired evolutionary algorithms

In the initial literature which introduced the QIGA, a binary representation was adopted, wherein each quantum chromosome was restricted to consist of a series of 0s and 1s. The methodology was modified to include real-valued vectors by da Cruz et al., [9]. As with binary-representation QIGA, real-valued QIGA maintains a distinction between a quantum population and an observed population of, in this case, real-valued solution vectors. However the quantum individuals have a different form to those in binary-representation QIGA. The quantum population $Q(t)$ is comprised of $N$ quantum individuals $(q_i : i = 1, 2, 3, \ldots, N)$, where each individual $i$ is comprised of $G$ genes $(g_{ij} : j = 1, 2, 3, \ldots, G)$. Each of these genes consist of a pair of values $q_{ij} = (p_{ij}, \sigma_{ij})$ where $p_{ij}, \sigma_{ij} \in \Re$ represent the mean and the width of a square pulse. Representing a gene in this manner has a parallel with the quantum concept of superposition of states as a gene is specified by a range of possible values, rather than by a single unique value.

The original QIGA algorithms, e.g., [12,14] are based very closely on physical qubits, but the "quantum-inspired" algorithm of da Cruz et al. [9] used in this chapter draws less inspiration from quantum mechanics since it:

- does not use the idea of a quantum system (in particular, no qubits);
- only allows for constructive (not destructive) interference, and that interference is among "wave-functions" of *different* individuals;
- uses real numbers as weights, rather than the complex numbers which arise in superposition of states in physical systems;
- the PDFs used (uniform distributions) are not those arising in physical systems.

However, the da Cruz et al algorithm does periodically sample from a distribution to get a "classical" population, which can be regarded as a wave-function (quantum state) collapsing to a classical state upon observation.

#### Algorithm

The real-valued QIGA algorithm is as follows

```
Set t=0

Initialise Q(t) of N individuals with G genes

While (t < max t)
    Create the PDFs (and corresponding CDFs, which describe the probability
        distributions of real-valued random variables, see equation(6)) for
        each gene locus using the quantum individuals
    Create a temporary population, denoted E(T), of K real-valued solution
        vectors by observing Q(t) (via the CDFs)

    If (t=0) Then C(t)=E(t)
    (Note: the population C(T) is maintained between iterations of the algorithm)
    Else    E(t)=Outcome of crossover between E(t) and C(t)
            Evaluate E(t)
            C(t)= K best individuals from E(t) U C(t)
```

```
    End if

    With the N best individuals from C(t)
    Q(t+1)=Output of translate operation on Q(t)
    Q(t+1)=Output of resize operation on Q(t+1)
    t=t+1
Endwhile
```

## Initialising the Quantum Population

A *quantum chromosome*, which is observed to give a specific solution vector of real-numbers, is made up of several quantum genes. The number of genes is determined by the required dimensionality of the solution vector. At the start of the algorithm, each quantum gene is initialised by randomly selecting a value from within the range of allowable values for that dimension. A gene's width value is set to the range of allowable values for the dimension. For example, if the known allowable values for dimension $j$ are $[-75, 75]$ then $q_{ij}$ (dimension $j$ in quantum chromosome $i$) is initially determined by randomly selecting a value from this range (say) -50. The corresponding width value will be 150. Hence, $q_{ij} = (-50, 150)$. The square pulse need not be entirely within the allowable range for a dimension when it is initially created as the algorithm will automatically adjust for this as it executes. The height of the pulse arising from a gene $j$ in chromosome $i$ is calculated using

$$h_{ij} = \frac{1/\sigma_{ij}}{N} \tag{7.4}$$

where $N$ is the number of individuals in the quantum population. This equation ensures that the probability density functions (PDFs) (see next subsection) used to generate the observed individual solution vectors will have a total area equal to one. Fig. 7.2 provides an illustration of a quantum gene where N=4.

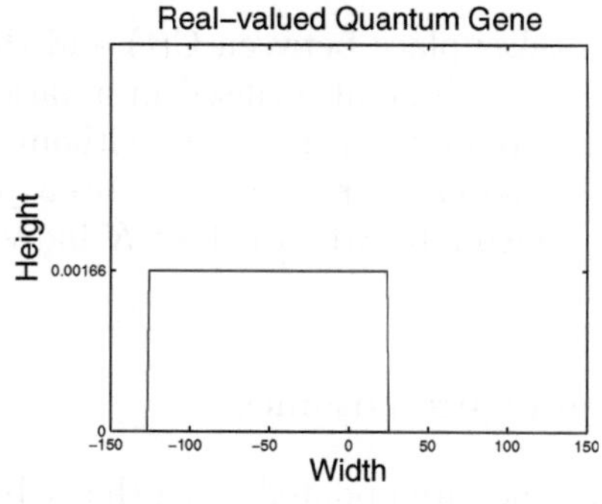


**Fig. 7.2.** A square pulse, representing a quantum gene, with a width of 150, centred on -50. The height of the pulse is $0.00166\dot{6}$

### Observing the Quantum Chromosomes

In order to generate a population of real-valued solution vectors, a series of observations must be undertaken using the population of quantum chromosomes (individuals). A pseudo-interference process between the quantum individuals is simulated by summing up the square pulses for each individual gene across all members of the quantum population. This generates a separate PDF (just the sum of the square pulses) for each gene and eq. 7.4 ensures that the area under this PDF is one. Hence, the PDF for gene $j$ on iteration $t$ is

$$PDF_j(t) = \sum_{i}^{j} g_{ij} \tag{7.5}$$

where $g_{ij}$ is the square pulse of the $j^{th}$ gene of the $i^{th}$ quantum individual (of $N$). To use this information to obtain an observation, the PDF is first converted into its corresponding Cumulative Distribution Function (CDF)

$$CDF_j(x) = \int_{L_j}^{U_j} PDF_j(x)dx \tag{7.6}$$

where $U_j$ and $L_j$ are the upper and lower limits of the probability distribution. By generating a random number $r$ from (0,1), the CDF can be used to obtain an observation of a real number $x$, where $x = CDF^{-1}(r)$. If the generated value $x$ is outside the allowable real valued range for that dimension, the generated value is limited to its allowable boundary value. A separate PDF and CDF is calculated for each of the $G$ gene positions. Once these have been calculated, the observation process is iterated to create a temporary population with $K$ members, denoted E(t).

### Crossover Mechanism

The crossover operation takes place between C(t) and the temporary population E(t). This step could be operationalised in a variety of ways with [9] choosing to adopt a variant of uniform crossover, without an explicit selection operator. After the $K$ crossover operations have been performed, with the resulting children being copied into E(t), the best $K$ individuals $\in C(t) \cup E(t)$ are copied into C(t).

### Updating the Quantum Chromosomes

The $N$ quantum chromosomes are updated using the $N$ best individuals from C(t) after performing the crossover step. Each quantum gene's mean value is altered using

$$p_{ij} = c_{ij} \tag{7.7}$$

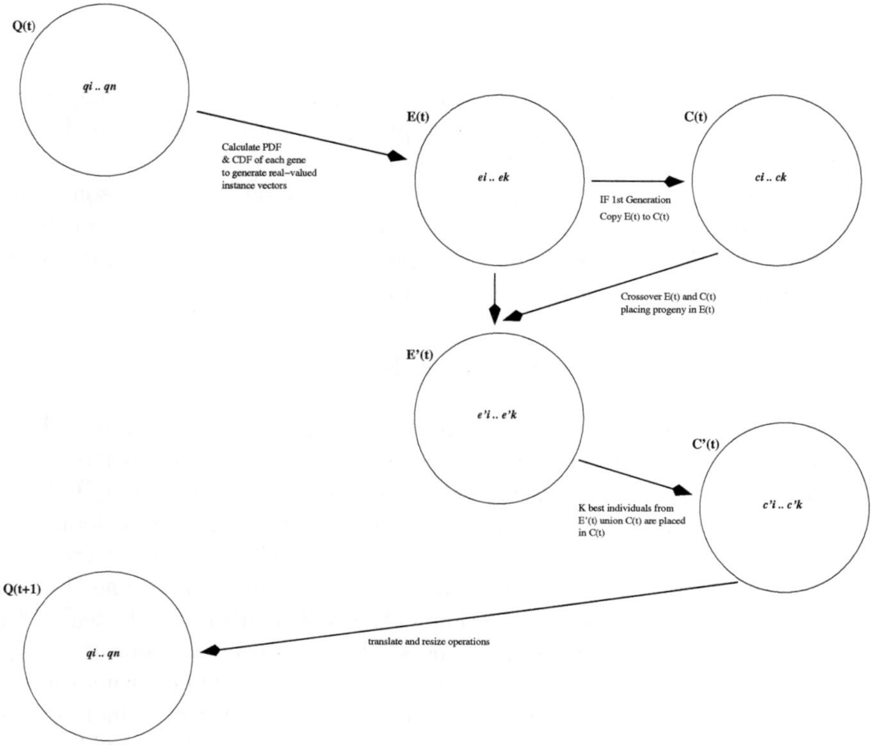


**Fig. 7.3.** An illustration of the process in the creation of generation t+1 from t.

so that the mean value of the $j^{th}$ gene of the $i^{th}$ quantum chromosome is given by the corresponding $j^{th}$ value of the $i^{th}$ ranked individual in $C(t)$.

The next step is to update the corresponding width value of the $j^{th}$ gene. The objective of this process is to vary the exploration / exploitation characteristics of the search algorithm, depending on the feedback from previous iterations. If the search process is continuing to uncover many new better solutions, then the exploration phase should be continued by keeping the widths relatively broad. However, if the search process is not uncovering many new better solutions, the widths are reduced in order to encourage finer-grained search around already discovered good regions of the solution space. There are multiple ways this general approach could be operationalised. For example, [9] suggests use of the 1/5th mutation rule from Evolutionary Strategies [24] whereby

$$if\ \phi < 1/5 \text{ then } \sigma_{ij} = \sigma_{ij} g$$

$$if\ \phi > 1/5 \text{ then } \sigma_{ij} = \sigma_{ij} / g$$

$$if\ \phi = 1/5 \text{ then } \sigma_{ij} = \sigma_{ij}$$

where $\sigma_{ij}$ is the width of the $i^{th}$ quantum chromosome's $j^{th}$ gene, $g$ is a constant in the range $[0,1]$ and $\phi$ is the proportion of individuals in the new population that have improved their fitness.

In this study we update the width of the $i^{th}$ quantum chromosome's $j^{th}$ gene by comparing each successive generation's best fitness function. If the best fitness function has improved (disimproved) we shrink (enlarge) the width in order to improve the local (global) search.

### QIGA vs Canonical Genetic Algorithm

A number of distinctions between the QIGA above and the canonical GA (CGA) can be noted. In the CGA, the population of solutions persists from generation to generation, albeit in a changing form. In contrast, in QIGA, the population of solutions in $P(t)$ are discarded at the end of each loop. The described QIGA, unlike CGA, does not have explicit concepts of crossover or mutation. However, the adaptation of the quantum chromosomes in each iteration does embed implicit selection as the best solution is selected and is used to adapt the quantum chromosome(s). The crossover and mutation steps are also implicitly present, as the adaptation of the quantum chromosome in effect creates diversity, as it makes different states of the system more or less likely over time. Another distinction between the QIGA and the CGA is that the CGA operates directly on representations of the solution (the members of the current population of solutions), whereas in QIGA the update step is performed on the probability amplitudes of the ground states for each qubit making up the quantum chromosome(s).

## 7.3 Option Pricing Model Calibration

An optimisation problem in financial modelling is considered to test the performance of the QIGA. The optimisation involves calibrating an option pricing model to observed market data. Calibration is a method of choosing model parameters so that the distance between a set of model option prices and market option prices is minimised, where distance is some metric such as the sum of squared errors or the sum of squared percentage errors. The parameters can be thought to resemble the market's view on current option prices and the underlying asset price. In calibration we do not explicitly take into account any historical data. All necessary information is contained in today's option prices which can be observed in the market. Practitioners frequently calibrate option pricing models so that the models provide reasonable fits to current observed market option prices and they then use these models to price exotic derivatives or for hedging purposes. In this paper we calibrate a very recent extension of the *Variance Gamma* option pricing model [20–22] known as the *Variance Gamma Scaled Self-Decomposable* (*VGSSD*) model [5] to FTSE 100 index option data.

A European call (put) option on an asset $S_t$ with maturity date $T$ and strike price $K$ is defined as a contingent claim with payoff at time $T$ given by $\max[S_T - K, 0]$ ($\max[K - S_T, 0]$). The well known Black-Scholes (BS) formula for the price of a call at time $t$ on this asset is given by

$$C_{BS}(S_t, K, r, q, \tau; \sigma) = S_t e^{-q\tau} N(d_1) - K e^{-r\tau} N(d_1) \tag{7.8}$$

$$d_1 = \frac{-\ln m + \left(r - q + \frac{1}{2}\sigma^2\right)\tau}{\sigma\sqrt{\tau}} \qquad d_2 = d_1 - \sigma\sqrt{\tau} \tag{7.9}$$

where $\tau = T - t$ is the time-to-maturity, $t$ is the current time, $m = K/S$ is the moneyness of the option, $r$ and $q$ are the continuously compounded risk-free rate and dividend yield and $N(\cdot)$ is the cumulative normal distribution function. Suppose a market option price, denoted by $C_M(S_t, K)$, is observed. The Black-Scholes implied volatility for this option price is that value of volatility which equates the BS model price to the market option price as follows

$$\begin{aligned} \sigma_{BS}(S_t, K) &> 0 \\ C_{BS}(S_t, K, r, \tau; \sigma_{BS}(S_t, K)) &= C_M(S_t, K) \end{aligned} \tag{7.10}$$

If the assumptions underlying the BS option pricing model were correct, the BS implied volatilities for options on the same underlying asset would be constant for different strike prices and maturities. However in reality the BS implied volatilities are varying over strike price and maturity. Given that the options are written on a single underlying asset this result seems at first paradoxical, i.e. we have a number of different implied volatilities for a single asset which should only have one measure for its volatility. Yet if we relax some of the assumptions in the BS model, such as allowing for a more complex data generating process for the asset price than the log normal stochastic process (as assumed by BS), and take into account the resulting complications, this result begins to make sense and is simply highlighting the erroneous assumptions that underpin the BS model.

Many different option pricing models have been proposed as alternatives to the BS model. Examples include stochastic volatility models and jump diffusion models which allow for more complex asset price dynamics. We examine a simple extension of a very popular option pricing model known as the Variance Gamma (*VG*) option pricing model. The extension of the model is called the Variance Gamma Scaled Self-Decomposable (*VGSSD*) model. The idea of the *VG* process is to model the continuously compounded returns of the stock price occurring on business time rather than on calendar time using a time transformation of a Brownian motion. The resulting model is a three parameter model where roughly speaking we can interpret the parameters as controlling volatility, skewness and kurtosis, denoted respectively as $\sigma, \theta$ and $\nu$, of the underlying asset returns distribution. Closed form option pricing formulae exist under the $VG$ model [22]. The model performs well at fitting a range of options prices with different strike prices at one maturity but fails to fit option prices at several maturities. This is because the returns are independent in the $VG$ model and this results in skewness and excess kurtosis of the $VG$ returns density function approaching zero too quickly as maturity increases. This resulted in [5] proposing a number of alternative models, one of which is based on the $VG$ model. The model proposes that a *VGSSD* random variable at time $\tau$ has the same distribution as a $VG$ random variable at unit time multiplied by $\tau^\gamma$, where $\gamma$ is an additional parameter of the model that induces volatility clustering in the returns. The resulting random process has a scaled density function, i.e. skewness and excess kurtosis remain constant as time-to-maturity increases. The *VGSSD* model performs much better than the $VG$ model at calibrating to a range of options across both strike price and maturity at the expense of one additional parameter. The characteristic function (Fourier transform of its density function) of the *VGSSD* process, $X_\tau$, has a very simple form and is given by

$$\phi_{X_\tau}(u) = E[\exp(\mathrm{i}uX_\tau)] = \left(\frac{1}{1 - \mathrm{i}u\nu\theta\tau^\gamma + \frac{1}{2}u^2\nu\sigma^2\tau^{2\gamma}}\right)^{\frac{1}{\nu}} \tag{7.11}$$

where $u$ is a Fourier transform parameter and i is the imaginary number with $\mathrm{i} = \sqrt{-1}$. The characteristic function of the logarithm of the stock price is given by

$$\phi_{\ln S_T}(u) = \exp\{\mathrm{i}u\,(\ln S_t + (r-q)\tau - \ln\phi_{X_\tau}(-\mathrm{i}))\}\,\phi_{X_\tau}(u) \tag{7.12}$$

This form ensures that the expectation of the future stock price in the risk neutral world is given by $E[S_T] = S_t \exp((r-q)\,\tau)$ where we recall that $\tau = T - t$. Option prices can be computed using the well known fast Fourier transform approach of [4]. A call option price with strike price $K$ and time-to-maturity $\tau$ is given by

$$c(K,\tau) = \frac{\exp(-\alpha\ln(K))}{\pi}\int_0^{+\infty}\exp(-\mathrm{i}v\ln(K))\,\psi_\tau(v)\,\mathrm{d}v \tag{7.13}$$

where

$$\psi_\tau(v) = \frac{\exp(-r\tau)\,\phi_{\ln S_T}(v-(\alpha+1)\,\mathrm{i})}{\alpha^2+\alpha-v^2+\mathrm{i}\,(2\alpha+1)\,v} \tag{7.14}$$

$\alpha$ is a dampening parameter that must be included for numerical reasons, it is set to $\alpha = 4$ in this analysis, and the integral in equation 7.13 is computed numerically using a fast Fourier transform (FFT), see [4] for more details.

## 7.4 Experimental Approach

Market makers in the options markets quote BS implied volatilities rather than option prices even though they realise BS is a flawed model. Table 7.1 depicts end-of-day settlement Black-Scholes implied volatilities for FTSE 100 European options on the 17 March 2006 for different strike prices and time-to-maturities. As can be seen the BS implied volatilities are not constant across the strike price and the maturity date. These implied volatilities are converted into market call and put prices by substituting the BS implied volatilities into the Black-Scholes formula. The following input parameters were used to calculate the option prices, the index price is the FTSE 100 index itself $S_t = 5999.4$, the interest rate is the one month *Libor* rate converted into a continuously compounded rate $r = 0.0452$ and the dividend yield is a continuously compounded dividend yield downloaded from *Datastream* and is $q = 0.0306$. These prices are then taken to be the observed market option prices. Out-of-the money (OTM) put prices were used for $K < S$ and OTM call prices were used for $K > S$ in the calibration. The calibration problem now amounts to choosing an optimum parameter vector $\Theta = \{\sigma, \nu, \theta, \gamma\}$ such that an objective function $G(\Theta)$ is minimised. In this paper the objective function is chosen to be the absolute average percentage error (APE)

$$\mathrm{G}(\Theta) = \frac{1}{N}\sum_{i=1}^{N}\left|\frac{C_i - C_i(\Theta)}{C_i}\right|$$

where $C_i$ is the observed market price on the $i$-th option (could be a call or a put) and $C_i(\Theta)$ is the *VGSSD* model price of the $i$-th option with parameter vector $\Theta$. One of the difficulties in model calibration is that the available market information may be insufficient to completely identify the parameters of a model [6]. If the model is sufficiently rich relative to the number of market prices available, a number of possible parameter vector combinations will be compatible with market prices and the objective function $G(\Theta)$ may not be convex function of $\Theta$. A plot of the objective function versus the two parameters controlling skewness and kurtosis of the asset returns distribution, $\theta$ and $\nu$, whilst keeping $\sigma$ and $\gamma$ fixed at $\sigma = 0.1175$ and $\gamma = 0.5802$ is shown in figure 7-4(a). Although the objective function looks well behaved other studies show that there are some potential problems. A graph of a very similar objective function is plotted in [6], for the *VG* option pricing model, using DAX index option data and the potential for gradient based optimisers to converge to a local rather than the global minimum is illustrated due to the fact that certain parameters have off setting effects in the *VG* model. Options are priced using the *VG* model with FFTs by [15] who outlines the potential for option prices to explode to infinity for certain parameter values of the *VG* model where conditions on the parameter values that keep the characteristic function finite do not hold. These potential problems provide the motivation to use an evolutionary based optimiser for calibrating the *VGSSD* model.

**Table 7.1.** Market BS implied volatilities (%) for FTSE 100 index options on the 17 March 2006. The strike prices and maturities (in years) are given in the table and the other observable inputs are $S = 5999.4, r = 0.0452$ and $q = 0.0306$.

| | Strike Price | | | | | | | | | | |
|---|---|---|---|---|---|---|---|---|---|---|---|
| Maturity | 4196.5 | 4796 | 5395.5 | 5695.2 | 5845.1 | 5995 | 6144.9 | 6294.7 | 6594 | 7194 | 7793.5 |
| 0.0959 | | | | 13.76 | 12.41 | 11.13 | 10.44 | 10.94 | | | |
| 0.1726 | | | 15.43 | 13.27 | 12.20 | 11.14 | 10.39 | 10.32 | | | |
| 0.2493 | | | 15.21 | 13.28 | 12.32 | 11.37 | 10.59 | 10.22 | 10.26 | | |
| 0.4986 | | 18.43 | 15.22 | 13.65 | 12.87 | 12.10 | 11.42 | 10.98 | 10.76 | | |
| 0.7479 | 21.24 | 18.28 | 15.45 | 14.07 | 13.38 | 12.70 | 12.06 | 11.55 | 10.96 | 10.76 | |
| 0.9973 | 20.98 | 18.27 | 15.69 | 14.42 | 13.80 | 13.18 | 12.57 | 12.06 | 11.35 | 10.85 | |
| 1.2466 | 20.85 | 18.33 | 15.94 | 14.76 | 14.18 | 13.60 | 13.04 | 12.54 | 11.79 | 11.09 | 10.80 |

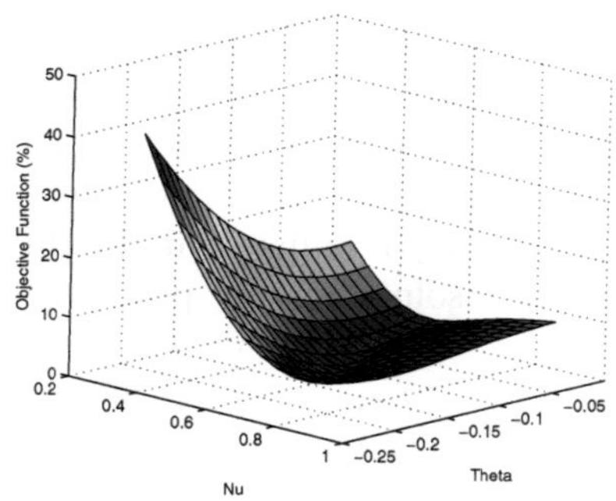


(a) Objective function vs parameters

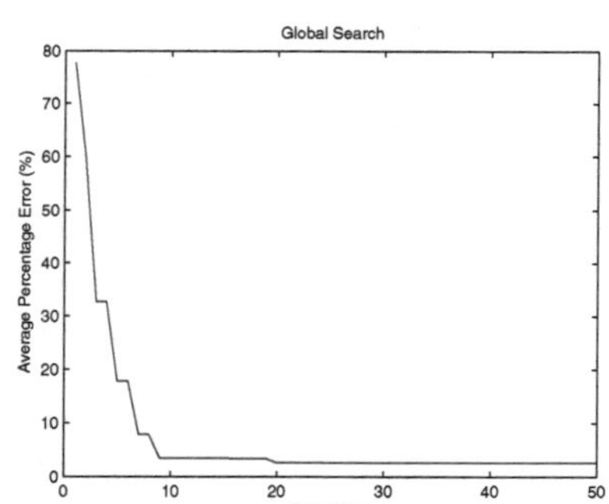


(b) Objective function vs generation no.

**Fig. 7.4.** Objective function versus model parameters $\nu$ and $\theta$ and objective function versus generation number.

## 7.5 Results

In all runs of the QIGA, a population size of 20 observed chromosomes was used, the algorithm was allowed to run for 30 generations, and all reported results are averaged over 30 runs. In order to provide a benchmark for the results obtained by the QIGA a deterministic Matlab optimiser called *fminsearch* was run 30 times with different initial parameter vectors. *Fminsearch*, multidimensional unconstrained nonlinear minimization, uses the Nelder-Mead simplex search method. This nonlinear optimization algorithm approximately finds a locally optimal solution to a problem with variables when the objective function varies smoothly, see Lagarias et al [17]. It is a direct search method that does not use numerical or analytic gradients. The optimiser converged to different values for the parameter vector $\Theta$ for different initialisations of $\Theta$. The parameter vector associated with the optimal value for the objective function $G$ was chosen to compute the average percentage error (APE) and the model prices. Figure 7-4(b) depicts the evolution of the global objective function $G$, measured using average percentage error (APE), as a function of the generation number for a single run of the algorithm. Figures 7-5(a), 7-5(b), 7-5(c) and 7-5(d) depict the evolution of the parameters $\sigma, \nu, \theta$ and $\gamma$ as a function of the generation number for a single run of the algorithm.

The best results from the last generation of the algorithm, averaged over 30 runs, are reported in the first column Table 7.2. The average results from the last generation of the algorithm, averaged over 30 runs, are reported in the second column of Table 7.2. As can be seen in Table 7.2 the optimal APE is very low at 1.29% (this compares favourably to the optimal matlab value of 1.27%) and the parameter values are very close to the parameter values from the Matlab optimiser(*fminsearch*).

The results from the second column of Table 7.2 are the average result from the last generation of the algorithm, averaged over 30 runs. The APE is calculated by computing a single set of model prices from the average parameter vector $\Theta$. The APE is higher than the optimal values at 3.34% indicating that some of the runs converged too early to a suboptimal result. However the average parameter values are not that different from the optimal parameter values indicating that only a small number of the 30 runs gave poor results.

### Model vs Market Implied Volatilities

In order to give a more complete picture of the results figure 7.6 depicts market and model implied volatilities versus the moneyness of the option, where moneyess is equal to $\frac{K}{S}$, for all the options used in the analysis. The model implied volatilities are calculated from the model prices using the optimal parameter values from Table 7.2. Model and market implied volatilities are plotted as opposed to model and market option prices because implied volatilities depict the calibration performance of the option pricing model in a much clearer way than option prices alone. This is because implied volatilities vary far less with strike price and maturity than option prices themselves. The objective function depends on model option prices as opposed to model implied volatilities because optimising over implied volatilities would be very time consuming since implied volatility has to be calculated numerically given the option price and other input parameters. However, if the model option prices are very close to the market option prices, the model implied volatilities will also be very close to the market implied volatilities. Figure 7.6 outlines the success of the parsimonious *VGSSD* model in terms of calibrating to a wide range of option prices. The *VGSSD* model performs best at the shorter maturities but more flexibility would be needed to match the deep out-of-the-money put options at longer maturities. However for a four parameter model the calibration performance is still very promising.

### Sensitivity Analysis

In previous expositions of the real-valued QIGA, detailed sensitivity analysis results were not reported. In order to gain greater insight into the operation of the algorithm, and to guide future applications of it, we undertook such an analysis by systematically investigating a variety of parameter settings for shrinkage, enlargement and crossover. The results of the optimal APE value as a function of the enlargement and shrink parameters are reported in table 7.3. The crossover rate is fixed at 0.3, a population size of 20 and a generation number of 50 are used. Figure 7-7(a) graphs these results.

The APE is less sensitive to the enlargement parameter than to the shrink parameter. The shrink parameter forces the algorithm to converge faster and this has a strong effect on the algorithms performance. The enlargement parameter causes the algorithm to widen the search space, however the crossover rate can also do this albeit using a different method, and this is why the algorithm is less sensitive to the enlargement parameter provided the crossover rate is a reasonable value. In this analysis the shrinkage parameter is set to 0.7 and the enlargement parameter is set to 1.2 as this provided a reasonable trade-off in the convergence speed of the algorithm versus the search space of the algorithm.

Table 7.4 reports the optimal APE value as a function of the shrink parameter and the crossover rate. The enlargement parameter is set to 1.2 and a population size of 20 and generation number of 50 are used. Figure 7-7(b) graphs these results. The algorithm performs more optimally with the crossover rate set to 0.3 and the shrink parameter set to 0.7. Low values of the crossover rate results in a smaller global search space and high values results in almost random search so intermediate values provide a reasonable trade-off between exploitation and exploration. Similarly the shrink parameter forces the algorithm to narrow the search space in the region of the current best solution, however this may be a local minima so the shrink parameter should not be made too small.

**Table 7.2.** Results of QIGA. The optimal and average parameter values from the last generation are averaged over 30 runs and compared with the parameter values from 30 runs of a Matlab optimiser.

| Parameter | QIGA Optima | Mean | Standard deviation | Matlab optimisation |
|---|---|---|---|---|
| $\sigma$ | 0.1181 | 0.1034 | 0.0618 | 0.1175 |
| $\nu$ | 0.5000 | 0.4088 | 0.1058 | 0.4975 |
| $\theta$ | -0.0941 | -0.1067 | 0.0993 | -0.1084 |
| $\gamma$ | 0.5662 | 0.5231 | 0.1017 | 0.5802 |
| APE (%) | 1.29 | 3.34 | 1.04 | 1.27 |

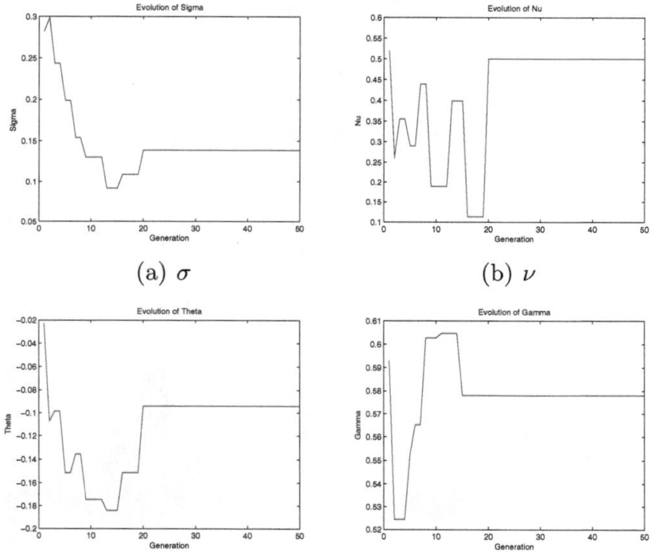


(a) $\sigma$ (b) $\nu$

(c) $\theta$ (d) $\gamma$

**Fig. 7.5.** Evolution of parameters over time.

**Table 7.3.** This table reports the APE (%) for different enlargement and shrinkage values.

| Enlarge\Shrink | 0.5 | 0.6 | 0.7 | 0.8 | 0.9 |
|---|---|---|---|---|---|
| 1.1 | 34.44 | 22.34 | 4.18 | 2.85 | 2.53 |
| 1.2 | 3.71 | 5.53 | 3.74 | 10.34 | 4.59 |
| 1.3 | 4.14 | 16.06 | 6.75 | 4.56 | 17.54 |
| 1.4 | 5.39 | 12.20 | 3.19 | 3.57 | 3.71 |
| 1.5 | 54.56 | 3.87 | 2.51 | 7.16 | 4.85 |

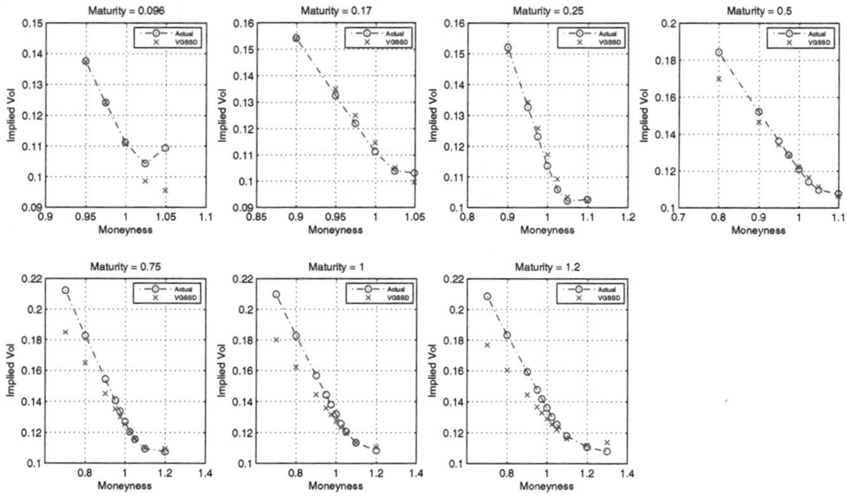


**Fig. 7.6.** Market and model implied volatilities.

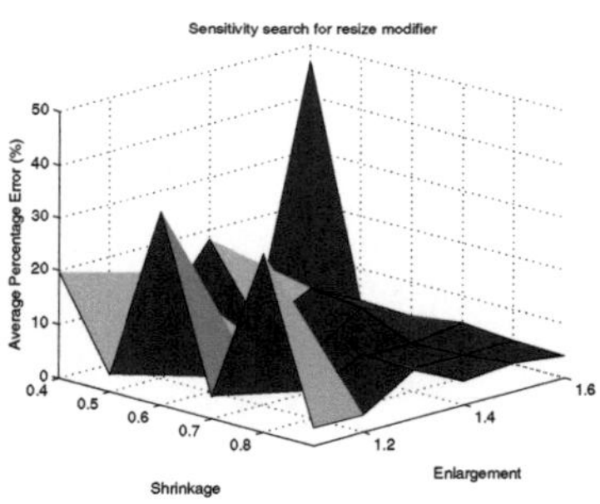


(a) sensitivity

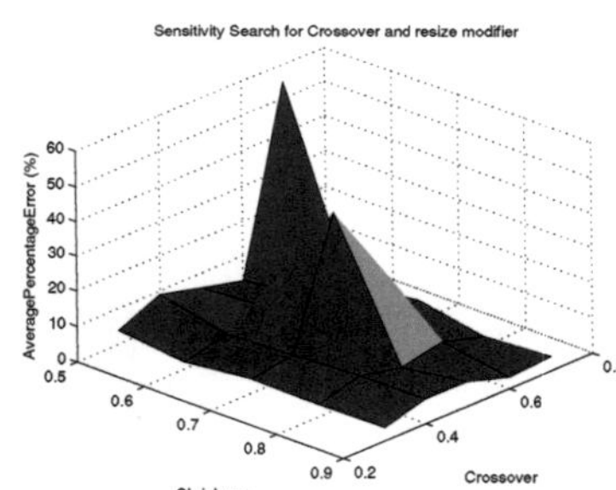


(b) shrinkage and crossover

**Fig. 7.7.** Sensitivity search

**Table 7.4.** This table reports the APE (%) for different shrinkage and crossover values.

| Crossover\Shrink | 0.5 | 0.6 | 0.7 | 0.8 | 0.9 |
|---|---|---|---|---|---|
| 0.3 | 3.84 | 1.90 | 4.30 | 3.68 | 3.85 |
| 0.4 | 8.96 | 4.74 | 5.82 | 6.82 | 7.30 |
| 0.5 | 5.81 | 7.20 | 41.39 | 4.17 | 7.09 |
| 0.6 | 2.81 | 5.84 | 3.39 | 6.30 | 3.32 |
| 0.7 | 54.85 | 3.29 | 6.25 | 2.72 | 4.08 |

## 7.6 Conclusions

The real-valued QIEA is a novel form of representation which could be hybridised with multiple real-valued search algorithms apart from the GA such as PSO or DE. This chapter illustrates how a quantum-inspired evolutionary algorithm can be constructed and examines the utility of the resulting algorithm on an important problem in financial modelling known as model calibration. The results from the algorithm are shown to be robust and comparable to those of other algorithms.

This underpins earlier proof of concept exploration studies using real valued quantum-inspired evolutionary algorithm. It is also noted that this paper reports the first application of a QIGA to the financial domain. Several extensions of the methodology in this study are indicated for future work. The first extension would be to extend the real-valued QIGA to a higher dimensional setting in order to examine, and highlight, the computational benefits of QIGA. The algorithm offers substantial potential for calibrating more complex financial models than the *VGSSD* option pricing model. Future work could also assess the benefits of operationalising the key steps of the QIGA in alternative ways, for example, by implementing alternative diversity-generating mechanisms in updating C(t) or by implementing alternative mechanisms for altering $\sigma$ during the algorithm.

## References


1. Benioff, P. (1980). The Computer as a Physical System: A Microscopic Quantum Mechanical Hamiltonian Model of Computers as Represented by Turing Machines. *Journal of Statistical Physics*, 22, 563–591.
2. Black, F. and Scholes, M. (1973). The pricing of options and corporate liabilities, *Journal of Political Economy*, 81:37–654.
3. Brabazon, A. and O'Neill, M. (2006). *Biologically-inspired Algorithms for Financial Modelling*, Berlin: Springer.
4. Carr, P. and Madan, D.P. (1999). Option valuation using the fast Fourier transform. *Journal of Business*, 75:305–332.
5. Carr, P., H. Geman, D. Madan and M. Yor (2007). Self decomposability and options pricing. *Mathematical Finance*, 17 (1).
6. Cont, R. and Ben Hamida, S. (2005). Recovering volatility from option prices by evolutionary optimisation, *Journal of Computational Finance*, 8 (4), Summer 2005.
7. da Cruz, A., Barbosa, C., Pacheco, M. and Vellasco, M. (2004) Quantum-Inspired Evolutionary Algorithms and Its Application to Numerical Optimization Problems, *ICONIP*, 2004, pp:212-217
8. da Cruz, A., Pacheco, M., Vellasco, M. and Barbosa, C. (2005) Cultural Operators for a Quantum-Inspired Evolutionary Algorithm Applied to Numerical Optimization Problems, *IWINAC*
9. da Cruz, A., Vellasco, M. and Pacheco, M. (2006). Quantum-inspired evolutionary algorithm for numerical optimization, in *Proceedings of the 2006*

*IEEE Congress on Evolutionary Computation (CEC 2006)*, 16-21 July, Vancouver, pp. 9180–9187, IEEE Press.

10. Feynman, R. (1982). Simulating Physics with Computers. *International Journal of Theoretical Physics*, 21 (6&7), 467–488.
11. Garavaglia, S. (2002). A quantum-inspired self-organizing map (QISOM), *Proceedings of 2002 International Joint Conference on Neural Networks* (IJCNN 2002),12-17 May 2002, pp. 1779–1784, IEEE Press.
12. Han, K-H. and Kim, J-H. (2002). Quantum-inspired evolutionary algorithm for a class of combinatorial optimization, *IEEE Transactions on Evolutionary Computation*, 6(6):580–593.
13. Han, K-H. and Kim, J-H. (2003). On setting the parameters of quantum-inspired evolutionary algorithm for practical applications, *Proceedings of IEEE Congress on Evolutionary Computing* (CEC 2003), 8 Aug-12 Dec. 2003, pp. 178–184, IEEE Press.
14. Han, K-H. and Kim, J-H. (2002). Quantum-inspired evolutionary algorithms with a new termination criterion, $H_\varepsilon$ gate and two-phase scheme, *IEEE Transactions on Evolutionary Computation*, 8(3):156–169.
15. Itkin, A. (2005). Pricing options with the *VG* model using FFTs. Working Paper. Moscow State Aviation University.
16. Jiao, L. and Li, Y. (2005). Quantum-inspired immune clonal optimization, *Proceedings of 2005 International Conference on Neural Networks and Brain* (ICNN&B 2005), 13-15 Oct. 2005, pp. 461–466, IEEE Press.
17. Lagarias, J.C., J. A. Reeds, M. H. Wright, and P. E. Wright. (1998) .Convergence Properties of the Nelder-Mead Simplex Method in Low Dimensions, *SIAM Journal of Optimization*, Vol. 9, Number 1, pp.112–147, 1998.
18. Lee C-D., Chen, Y-J., Huang, H-C., Hwang, R-C. and Yu, G-R. (2004). The non-stationary signal prediction by using quantum NN, *Proceedings of 2004 IEEE International Conference on Systems, Man and Cybernetics*, 10-13 Oct. 2002, pp. 3291–3295, IEEE Press.
19. Li, Y., Zhang, Y., Zhao, R. and Jiao, L. (2004). The immune quantum-inspired evolutionary algorithm, *Proceedings of 2004 IEEE International Conference on Systems, Man and Cybernetics*, 10-13 Oct. 2002, pp. 3301–3305, IEEE Press.
20. Madan, D. and Seneta, E. (1990). The VG model for share market returns, *Journal of Business*, 63:511–524.
21. Madan, D. and Milne, F. (1991). Option pricing with VG martingale components, *Mathematical Finance*, 1(4):39–55.
22. Madan, D., Carr, P. and Chang, E. (1998). The Variance Gamma Process and Option Pricing, *European Finance Review*, 2:79–105.
23. Narayanan, A. and Moore, M. (1996). Quantum-inspired genetic algorithms, *Proceedings of IEEE International Conference on Evolutionary Computation*, May 1996, pp. 61–66, IEEE Press.
24. Rechenberg, I. (1973). *Evolutionsstrategie: Optimierung Technisher Systeme nach Prinzipien der Biologischen Evolution*, Fromman-Holzboog Verlag, Stuggart.
25. Spector, L. (2004). *Automatic Quantum Computer Programming: A Genetic Programming Approach*, Boston, MA: Kluwer Academic Publishers.
26. Tsai, X-Y., Chen, Y-J., Huang, H-C., Chuang, S-J. and Hwang, R-C. (2005). Quantum NN vs NN in Signal Recognition, *Proceedings of the*

*Third International Conference on Information Technology and Applications* (ICITA 05), 4-7 July 2005, pp. 308–312, IEEE Press.

27. Yang, S., Wang, M. and Jiao, L. (2004). A genetic algorithm based on quantum chromosome, *Proceedings of IEEE International Conference on Signal Processing* (ICSP 04), 31 Aug- 4 Sept. 2004, pp. 1622–1625, IEEE Press.
28. Yang, S., Wang, M. and Jiao, L. (2004). A novel quantum evolutionary algorithm and its application, *Proceedings of IEEE Congress on Evolutionary Computation 2004* (CEC 2004), 19-23 June 2004, pp. 820–826, IEEE Press.
29. Yang, S., Wang, M. and Jiao, L. (2004). A Quantum Particle Swarm Optimization, in *Proceedings of the Congress on Evolutionary Computation 2004*, 1:320–324, New Jersey: IEEE Press.

# Author Index

MIX
Papier aus verantwortungsvollen Quellen
Paper from responsible sources
FSC® C105338

If you have any concerns about our products,
you can contact us on
**ProductSafety@springernature.com**

In case Publisher is established outside the EU,
the EU authorized representative is:
**Springer Nature Customer Service Center GmbH**
**Europaplatz 3, 69115 Heidelberg, Germany**

Printed by Libri Plureos GmbH
in Hamburg, Germany